A QUESTION OF STANDING

A QUESTION OF STANDING

THE HISTORY OF THE CIA

RHODRI JEFFREYS-JONES

OXFORD
UNIVERSITY PRESS

OXFORD
UNIVERSITY PRESS

Great Clarendon Street, Oxford, OX2 6DP,
United Kingdom

Oxford University Press is a department of the University of Oxford.
It furthers the University's objective of excellence in research, scholarship,
and education by publishing worldwide. Oxford is a registered trade mark of
Oxford University Press in the UK and in certain other countries

First Edition published in 2022

Impression: 1

Published in the United States of America by Oxford University Press
198 Madison Avenue, New York, NY 10016, United States of America

British Library Cataloguing in Publication Data
Data available

Library of Congress Control Number: 2021951468

ISBN 978–0–19–284796–6

DOI: 10.1093/oso/9780192847966.001.0001

Printed and bound in Canada

For Effie, George, and Sadie

Contents

Preface

Cold air blasted into the vestibule of the Harrington, cushioning the hotel's air-conditioned interior. Pushing outward on its doors I plunged into the August humidity. Some twenty minutes later I was at the Neptune saloon, close to the Capitol and the Library of Congress. The admiral's daughter had taken a break from dancing on the bar top. Clients relaxed. At least for a while she would not, as was her custom, leap unnervingly into an unsuspecting lap. Absent, too, was the barfly who claimed to be a nephew of the Louisiana populist Huey Long, and whose habit was to greet me with the unwelcome words, "Ah hate the British. Bah me a beer."

Richard Harris Smith sat nursing a Schlitz. I recognized him from a photo issued to publicize his book *OSS* (1972). That book was a gem. I was sympathetic to its excoriation of British imperial ways, though I did wonder if it presaged an American empire. Later on I got to know a little about Harris Smith. As a senior CIA officer, he had run the situation room established by the administration of Lyndon B. Johnson. He had been in charge there at the time of the questionable US military intervention in the Dominican Republic's civil war (1965).

Harris Smith was a charming, can-do kind of person. Learning of my interest in the history of the CIA and knowing that I lived in Scotland, he proposed that we organize a joint conference on the subject of the agency's past. This would be on the Isle of Skye. Overlooked by the Cuillin hills, we would plot a course leading to greater enlightenment. Perhaps Dame Flora would allow us to use Dunvegan Castle. Spurning the local, smoky whiskies, we would drink Speyside's Macallan single malt. He knew it was my favorite tipple and stated that, at CIA headquarters in Langley, VA, they drank nothing else.

The conference that would have brought back the good old days never came to pass. In the mid-1970s the American people and Congress rose up against the CIA's boondoggling practices. Harris Smith, a man I liked enormously, became the relic of a bygone age.[1]

The year 1975's intelligence investigations not only interrupted the flow of unvouchered funds but also led to other reforms such as more insistent congressional oversight. It was the first major upheaval in the CIA's history. The following chapters present an overview not only of upheavals but also of the agency's origins, successes, and failures. This is not with the aim of supplying comprehensive coverage. America is a world power and any attempt to cover the CIA's history in its entirety would be overambitious. Instead, in the style of an essayist I shall look at key events. Such events, for example, the Bay of Pigs episode, will be recognizable at least by name to students of the American past.

With the same object of making the CIA's history easier to the eye, I mainly avoid certain practices that have plagued the writing of intelligence history. They include the excessive use of acronyms. Further eschewed is the overemployment of sometimes obscure operational names. This book calls the Bay of Pigs invasion the Bay of Pigs invasion, in preference to Operation TRINIDAD, Operation ZAPATA, or Operation JMARC. Finally, I omit that page-defacing device, the uniform research locator, or URL.[2]

Observing the foregoing principles, we will expound the book's central thesis, that the agency's effectiveness has depended on its standing. The CIA's standing in the White House determined the scale of its impact. The agency struggled with other entities vying for position in the White House pecking order. Its success depended on the ability of a president to understand intelligence briefings but also on his willingness to heed them, a function of personality and political priorities. It depended not only on the CIA's performance but also on the communication skills of its leadership. Finally, in democratic America the standing of the CIA was discernibly dependent on public opinion.

The White House responds to public opinion because the president is an elected official. The president also heeds it indirectly because he is answerable to Congress. For members of the House and of the Senate, as for the chief executive, what shaped opinion mattered. Therefore, it mattered for the CIA. For the CIA needed appropriations, the prerogative of the House of Representatives, and it needed also Senate approval of its foreign-policy-affecting activities.

Domestic transgressions by the agency such as spying on students did not go down well with Congress. What the CIA did overseas mattered less. The CIA's standing abroad is still significant, though. As one observer after another affirmed, the agency's covert actions were the greatest single cause of anti-Americanism in the post-World War II era.

Those who mandated the covert operators to engage in excesses seriously damaged America's claim to be the champion of democracy. Great swathes of the world's population wanted a judicious blend of democracy, socialism, and capitalism, which was just what the United States itself had, but would not fully admit.[3] The CIA itself was a democratically created and overseen institution. But those who guided the CIA insisted, in their designs for foreign lands, on defining and enforcing a binary choice between socialism and democracy-plus-capitalism. Thus it came to pass that the agency contributed to the overthrow of democratically elected governments by covert but only too discoverable means. In this way, the CIA eroded America's international capabilities in the vital realm of "soft power." And when things came unstuck, as they regularly did, the agency suffered a loss of status in the White House that affected the credibility not just of its operators but of their colleagues the analysts as well.

To turn to its *raison d'être*, the CIA was in good measure the creation of World War II veterans who had served with the Office of Strategic Services (OSS). In a close examination of the agency's origins, we will further look at ways in which the more distant past helped to shape its mission. Well before the communist threat from the Soviet Union, U-1, an organism that grew out of World War I, reflected the idea that America needed peacetime central intelligence. Thus, there pre-existed a rationale for the continuation of central intelligence in the 1990s, when the demise of European communism seemed to make the CIA redundant.

That train of thought helped the CIA survive in the post-Cold War years until the second great upheaval in its existence, the Intelligence Reform and Terrorism Prevention Act of 2004. In the final section of the book, we examine the reasons for that reform, together with the reform's impact on the CIA and on US intelligence performance. The legislation demoted the CIA. The agency's standing suffered further when it gained the reputation of being a machine dedicated merely to torture and assassination. Yet was that reputation deserved? There were signs of improved intelligence analysis, an improvement that was in line with a tradition that had evolved ever since the days of U-1, and which survived even the turbulence of the Trump presidency.

Heartfelt thanks go to Richard Aldrich, Michael Donovan, Huw Dylan, Thomas Fingar, Christopher Fuller, John Ghazvinian, Gerry Hughes, Richard Immerman, Christopher Kojm, Ian Manners, Peter Reilly, Priscilla Roberts, Len Scott, Gregory Treverton, Hugh Wilford, and Randall Woods for their help in researching and thinking about the history of the CIA. I am grateful, also, to Brown University's Charlotte Scott and her fellow student-editors, for constructive criticism of drafts of my prefiguring article "American Espionage: Lessons from the Past," published in *The Brown Journal of World Affairs* (26/1, Fall/Winter 2019: 18–29). My editor at Oxford University Press, Matthew Cotton, steered the project through at a difficult time, and I am as ever appreciative of his dedication and skill.

I keep promising my wife Mary that I am working on my last book. My good fortune is that she matches her scepticism with loving support. I reciprocate her love. This volume is devoted to Mary's courageous daughter Effie, and to Effie's children George and Sadie.

I

The Road to U-1

The CIA is an American institution. To understand how the agency has fared in its seventy-five-year history, it is wise first to consider its precursors from the earliest days of the nation to the twentieth century. For while there was never anything quite like the agency before 1947, the manner of its creation as well as its subsequent fate reflected habits acquired and lessons learned over the seventeen preceding decades.

The CIA was created as an executive agency, and the issue of executive secret powers has a pedigree stretching back to the days of the Founding Fathers. In the Republic's early years, foreign threats to its very existence were so acute that Congress took a step back from democracy and invested implicit trust in President George Washington. In 1790, the legislators voted in favor of a "Contingent Fund of Foreign Intercourse" to enable Washington to pay his spies, allotting $40,000 in the first instance. Within three years, the "contingency" or "secret" fund had grown to $1 million, 12 percent of the federal budget.[1]

Washington's successors in the White House continued to pay spies from the Contingency Fund. They reimbursed them as individuals. Then in 1865 President Lincoln established the nation's first spy *agency*. In the course of the Civil War, he had so far experienced mixed fortunes with spies. When he engaged the services of Allan Pinkerton, the private detective proved more adept at boosting his own fame than estimating the strength of Confederate forces. In the cabinet meeting before his assassination, Lincoln finally established a federal agency, the United States Secret Service. He did so by executive order and without consulting Congress, safe in the knowledge that the Union was still at war and that his judgment would have to be trusted.

At first, with inflation threatening the Union's finances, the Secret Service tracked down greenback counterfeiters. When peace arrived, it switched its attention to southern moonshiners. What brought it into the realm of political controversy was a new assignment it received in 1871 from the "Radical Reconstructionists." In a significant precedent, America's spies would be put to work to champion a liberal cause. The Radicals in Congress wanted to deliver in full the newly conferred rights of millions of newly liberated African Americans. Happy to oblige them, Attorney General Amos T. Akerman tasked the Service's agents to track down the Ku Klux Klan. The Klan's elusive nightriders were using brutal methods to turn back the clock of southern history and enforce servitude. Former slave Brawley Gilmore of Union, South Carolina, remembered how heavily armed Klan horsemen would "come along at night riding de niggers like goats" and lining them up on the handrail of Turk Creek's bridge for target practice.

When the Secret Service returned to their familiar southern turf and asked for information in South Carolina, the freedman John Good saw his opportunity. He knew who the nightriders in his district were and that he could prove it. For Good was the local blacksmith, and he shod the very horses the Klan rode to terrify his kinsfolk. This gave him the opportunity to supply the Secret Service with proof. He marked the horses' hooves with bent nails so that the federal detectives could track them. The Klan worked that one out and killed John Good, but by that time he had already accomplished his mission.[2]

With the aid of such courageous individuals and of the United States Army, which still occupied swathes of the defeated South, the Secret Service crushed the Klan. Its victory was short-lived. By the end of the 1870s, the former Confederate states were back in the Union. Racism and anti-federalism received a new lease on life. Striking a chord that would resonate loudly and repeatedly into the era of the CIA, defenders of old southern ways accused the Secret Service of being an instrument of federal oppression. Others joined in. Lodging a procedural objection to the Service's budgetary appropriation, Republican Congressman Ebenezer B. Finley of Ohio demanded that the "secret detectives" should limit their work to that originally intended, the prevention of counterfeiting. Voicing a particularly sensitive charge, Finley accused the Secret Service of spying on his congressional colleagues. For such dubious reasons, the consensus behind the Secret Service began to disintegrate.[3]

Opposition to the instruments of clandestine power was particularly evident in the year 1893. This time, the opposition came from liberal, not conservative quarters, and such switches were to be common in the years to come. The Anti-Pinkerton Act aimed to curb the misdeeds of private detectives who impeded the activities of labor unions. Though rarely, if ever enforced, it reflected a significant segment of opinion. American anti-imperialists also took steps to frustrate clandestinity. The Republican Senator George F. Hoar was prominent among those who suspected that President Grover Cleveland was using secret methods to create an American empire in which non-white inhabitants would not enjoy the rights of US citizenship. When the Cleveland administration was negotiating the annexation of Hawaii in 1893, the Massachusetts senator secured passage of a resolution that came to be known as the Hoar Amendment. This allowed the president to employ a "mere agent" or "spy" to aid him in his duties, but forbade agents from conducting negotiations on behalf of the United States.[4] By the 1890s, then, the legitimacy and standing of American espionage had been challenged from the left as well as from the right.

The War of 1898 renewed national unity and helped to heal old Civil War wounds. The new-found unity meant widespread support for the Secret Service as it embarked on a counterespionage venture. At the beginning of the war, Spain's naval attaché Ramon de Carranza fled to Montreal, where he established a spy ring. Its objectives included identification of targets up and down the East Coast of the United States, for example, the financial district of Philadelphia, that might be bombarded by Spain's armored cruisers. Under the leadership of John E. Wilkie, the Secret Service investigated. On 7 May 1898 agents arrested George Downing in the act of mailing a letter about US Navy movements. Carranza described Downing as one of his "two best spies." Forty-eight hours later, Downing was found dead in his cell. Carranza conceded that he might have committed suicide but added, "or else they did it for him."[5]

At the time, the American press was too patriotic to pounce on that possibility. Overnight, Wilkie had become a person of standing—when he accepted an invitation to contribute to an anthology about the 1898 conflict, it was in his capacity as a "war leader." He had to answer no embarrassing questions about the methods of his Secret Service men.[6]

Patriotic fervor died down after the war's end, and with it the harmony over the Secret Service's activities. President Theodore Roosevelt reignited white southern antipathy. This was because he used the

Secret Service to investigate peonage, a new form of slavery the planters used to keep the black population in permanent debt and servitude. Roosevelt provoked further opposition when he set the Secret Service to investigate the fraudulent sale of federal timber lands in the Pacific West and discovered that the corruption reached all the way to Capitol Hill.

Not for the best of reasons, critics once again cried out that democracy was under threat. They fastened onto the fact that Roosevelt's attorney general, Charles Bonaparte, was the great-nephew of Napoleon Bonaparte. They likened his conduct to that of Napoleon's ruthless minister for police, Joseph Fouché. As for Wilkie, they dubbed him a dangerous bureaucratic empire builder. Some conservatives delighted in the appearance in 1905 of Thomas Dixon's novel *The Clansman*, which portrayed the Klan's nightriders as redeemers and condemned a thinly disguised Secret Service "with its secret factory of testimony and powers of tampering with verdicts."[7]

Roosevelt had his defenders. The *Pittsburg Leader* editorialized that the "virtuous congressmen" who accused the president of running a "political secret service" were hypocrites. They were "doing a bit of 'secret service' of their own. It is service for the trusts, the railroads, the timber thieves and others of that class."[8] Buoyed by such support, Roosevelt created a new bureau of investigation—after a number of name changes, this would be retitled in 1935 the Federal Bureau of Investigation (FBI).

The president took the precaution, however, of establishing the bureau by executive decree in a congressional adjournment. Not until 1947 would there be congressional approval of the creation of an intelligence-related agency.

The early "FBI" (as we shall call it for convenience's sake) enjoyed some early popularity and support when it attempted to enforce the provisions of the Mann Act of 1910, an attempt to suppress the "white slave trade," or prostitution. Then there was a real upsurge in support for the nation's investigative agencies when World War I broke out. Working closely with British intelligence even before America joined the war in 1917, the Secret Service pulled off some coups. For example, one day in 1915 the local New York German spy chief Dr. Heinrich F. Albert boarded an elevated train at 50th Street station, New York, carrying a briefcase containing compromising documents. Secret Service agent Frank Burke, a veteran of Wilkie's team of 1898, waited until Dr. Albert had nodded off to sleep, at which point he whisked away the briefcase. Franz von Papen, military attaché in Washington and the overall organizer of German spying in America, was dismayed. The future chancellor of Germany reportedly wrote to his wife, "Unfortunately they stole a fat suitcase from our good friend, Dr Albert, in the elevated. The English secret service, of course."[9]

The Secret Service had apparently fooled the Germans on this occasion, and it was generally successful in mopping up alien agents, as it had been in 1898. It basked in public approval. In future years, questions would be asked about the pro-British partisanship of the Woodrow Wilson administration, but by this time the Secret Service had retreated from politically freighted work. It had assumed a narrower responsibility and one that commanded universal approval, the protection of the president and his entourage. The FBI, which in the course of World War I muscled aside its older sister to assume a more important role, would in future be the greater source of controversy.

Meantime, the Wilson administration's quietest initiative had created a precedent for the CIA. President Wilson was a scholar who wanted to be informed about foreign relations. In September 1917 he would establish the Inquiry, a study group. The Inquiry's scholars used open sources to prepare databases that would guide peace negotiations. Earlier than that, though, there had been a more obscure initiative. Instead of establishing an agency, President Wilson's advisers acted with supreme secrecy in setting up a central intelligence unit within the State Department.

At its head was Frank L. Polk, who had become Counselor to the State Department in September 1915. When in July 1919 the counselor's title changed to "undersecretary of state," the clandestine unit assumed the title U-1, the term we shall use here for convenience. Its subdivisions were numbered U-2, U-3, and so on. Polk remained second in command at the Old Executive Building until he headed the American Commission to Negotiate Peace in 1919, and then became acting secretary of state. Thus, central intelligence spoke to power in the Woodrow Wilson years.

Unlike intelligence *agencies*, U-1 did not, with a few notable exceptions, send covert operatives into the field. Its task was to facilitate cooperation between existing agencies, for example, the Secret Service and military intelligence, to suggest and facilitate spying operations, and to communicate the results to the White House. In Eastern Europe, it oversaw activities that began to eclipse in that region the achievements of the UK's more celebrated spies. They included the undercover work of Emanuel Voska, who, in violation of the Hoar Amendment, facilitated self-determination within the old Austro-Hungarian Empire—he was an architect of the new state of Czechoslovakia, established in the fall of 1918.

America also assumed the role of paymaster for the intelligence activities of Voska's associate, W. Somerset Maugham. The English novelist had become estranged from the "exalted personages" who ran British intelligence, and his lover, F. Gerald Haxton, was an American who had been persecuted by the UK authorities.[10] Sir William Wiseman, the British officer who was responsible for Anglo-American intelligence liaison, persuaded Maugham to continue his work as a spy, with the result that Maugham travelled to St. Petersburg. He delivered to both London and his new masters in Washington high-quality intelligence on the Soviet takeover of Russia.

In the period of American neutrality, 1914–17, military attaché Franz von Papen oversaw espionage, sabotage of Allied assets in the United States, and attempts to recruit support for Germany's cause from Irish Americans and Indian Americans who sought independence for their countries of origin. U-1 oversaw successful efforts by the Secret Service to disrupt such enterprises.

The Department of State depended on trust and credibility in its dealings with foreign nations and generally eschewed direct espionage. However, even in the period of US neutrality, Secretary of State Robert Lansing took the risk of employing of agents directly. In April 1916 he created the Bureau of Secret Intelligence and placed it under the leadership of his colleague Leland Harrison. Lansing later explained that its purpose was "to issue instructions [to agents] and to digest and analyze their reports without their going through the regular channels of departmental correspondence."[11] The Bureau of Secret Intelligence ran its own agents; for example, it had a man in Cairo.

It was a limited enterprise. At the height of its operations, the bureau had no more than fifteen agents, and like U-1 it depended on the work of others. Later in 1916, Harrison did initiate a notable act of espionage. He guided the efforts of a team led by Secret Service agent Joseph M. Nye. This team of specialists tapped the telegraph and telephone lines of the German Embassy in Washington. Nye was able to alert Lansing to Germany's forthcoming announcement, on

31 January 1917, that it would resume unrestricted submarine warfare with no safeguarding against accidental attacks on neutral shipping. Between 3 February and 1 April, eight American ships were lost. Berlin had calculated that this might bring the United States into the war but that by then the stranglehold on Allied trade would have forced Britain and France to surrender.[12]

If U-boat atrocities contributed to the US entry into World War I, the Zimmermann affair was the more immediate precipitant. Here, U-1 had a significant presence. On 19 January 1917, Germany's foreign minister Arthur Zimmermann sent a secret diplomatic message proposing a military alliance between Germany and Mexico. If the United States entered the war and came out on the losing side, Mexico would recover the territory it had surrendered in the War of 1846–8, recovering Texas, Arizona, and New Mexico. The British intercepted the telegram, and an Admiralty unit known as Room 40 decoded it. Prime Minister Arthur Balfour delegated the handling of the issue to Room 40's director, Captain Reginald "Blinker" Hall. The man with the famous eye twitch decided to release the text to the United States authorities, but looked for a way to do so without alerting Germany to the fact that the UK had broken their codes.

Hall summoned his American confidant Edward "Ned" Bell. Second secretary at the US Embassy in London since 1913, Bell was a career diplomat in close touch with his colleagues at U-1. Hall wanted to demonstrate to Bell that the Zimmermann intercept was genuine, and he gave his friend an incomplete demonstration of how the original had been decoded—incomplete, because knowledge is power, and he did not want his US counterparts to be privy to Room 40's tricks of the trade. At the same time, he took the opportunity to explain to Bell that he wanted America to claim that it had independently obtained a copy of the telegram when it was forwarded within North America in a lower-level code. That was the course of action pursued. On 29 March Zimmermann admitted to the Reichstag that he had sent the telegram, and within days Germany and the United States were at war.

Bell and U-1 ran into the limits of Anglo-American intelligence cooperation. The Americans wanted not just the contents of the messages that Room 40 decoded but also the methodology and code-books that enabled UK decryption. Captain Hall held out. Bell reported to Harrison in January 1918 that the captain was "very shy at giving up more than he has to." He suggested that the "only way to manage these things is to establish an atmosphere of confidence."[13] The brash Army officer and cryptographer Herbert Yardley now tried after the more diplomatic Bell had failed. Yardley demanded access to Hall's secrets but found he was barred from visiting Room 40.

The time had come for a reassertion of American independence. Though valuing the British connection, U-1's leaders decided that they needed an intelligence capability of their own. In November 1917, Secretary of State Lansing promised Harrison "any financial assistance that might be required for this purpose." Harrison added, "It would seem desirable that the staff be selected with a view to keeping them on after the war."[14] In consequence, the State Department and US Army would jointly fund a cipher bureau that came to be known as the American Black Chamber. Herbert Yardley was the cipher bureau's first chief.

The chill emanating from Room 40 helped to stiffen American resolve. In May 1919, Bell wrote a critique of the British intelligence system with a view to considering how America should develop its own. He portrayed British intelligence as unprepared for the war, overstaffed, overfinanced, and chaotic.[15]

There has been some discussion as to whether U-1 should be considered to be a significant antecedent of the CIA. Thomas F. Troy, who served for some years as the CIA's official historian and who wrote on the agency's origins, disputed the idea.[16] More recently, however, the historian Mark Stout concluded that the CIA followed the "trail blazed by U-1" as a civilian intelligence agency.[17]

There were differences between U-1 and the CIA. U-1 ran only a small number of covert operations. The unit was not an independent agency but an integral part of the State Department. U-1 was directly involved in codebreaking and signals intelligence, activities mainly outside the remit of the CIA.[18] It was more successful than the later agency in guarding its secrets, including the secret of its existence. Indeed, its secrecy was such that, as Troy argued in his critique, the creators of the CIA in the 1940s may even have forgotten about its existence. Like the Secret Service and the FBI, U-1 was the creature of executive decree, and unlike the CIA it had no democratic mandate.

Similarities between U-1 and the CIA included the commitment to centralization, the duty of coordination of all sources of intelligence, the productive, yet competitive relationship with British intelligence, and a civilian identity. There was a further likeness, especially with the early CIA. U-1 was an Ivy League affair. Ned Bell was from "old New York" stock. He attended the Cutler preparatory school and Harvard, and belonged to exclusive societies such as the Racquet Club. Leland Harrison attended Eton and Harvard. Frank Polk was a Yale man. The whole U-1 coterie was Ivy League, and Woodrow Wilson, to whom they ultimately reported, had been president of Princeton.

The Ivy League factor contributed to mutual trust and an ability to keep secrets, and to U-1's high standing in the White House. At a time of wartime unity, the prospect of press probes or congressional investigation that might have diminished U-1's standing was minimal. But the war would not last forever, and the need for worldwide intelligence began to diminish as the 1919 peace negotiations wound down. Just like the CIA, U-1 would be susceptible to changes in circumstances and public mood. In U-1's case, those changes would prove fatal.

2

The OSS Model

Ray S. Cline attended Wiley High School in Terre Haute, Indiana, and won a scholarship to Harvard. In World War II, he served with the Office of Strategic Services (OSS), becoming head of current intelligence in 1944. After the war, he obtained a Ph.D. in History, also at Harvard. After successful completion of his dissertation on the US Army High Command in the recent war, Dr. Cline then joined the CIA.[1] He rose to become head of the agency's directorate of intelligence, 1962–6. Twenty years later, the historian/intelligence officer published a book reflecting on the CIA's present, past, and origins.

As befitted an instrument of a democracy, the CIA was a pluralistic institution whose personnel had differences of opinion on a number of issues.[2] But trends of thought still existed, and Cline's interpretation of the CIA's origins reflected views regularly held within the agency.

The Harvard graduate stated that the United States was "almost totally unprepared" for World War II. Pearl Harbor was a direct result of a failure of US intelligence coordination. Yet a sudden transformation then occurred. It was "almost a miracle that the United States built a creditable wartime record with its Office of Strategic Services, which served as a legacy for peacetime efforts, and then developed under the CIA during the 1950s and 1960s the best intelligence system in the world."

When the remedy came in the shape of the OSS, Dr. Cline added, it was not the result of lessons learned from the American past. A foreign power had come to the rescue. The UK had bestowed its "handsomest gift of all," the "concept of a central, coordinated intelligence system."[3]

The OSS veteran's views were mainstream.[4] They amounted to a powerfully backed myth—a myth devoted to a purpose. Until the creation of the CIA in 1947, the United States had not had a permanent

A Question of Standing: The History of the Cia. Rhodri Jeffreys-Jones. Oxford University Press.
© Rhodri Jeffreys-Jones 2022. DOI: 10.1093/oso/9780192847966.003.0002

agency devoted to central intelligence. To show that the nation would always need such a permanent and expensive capability, it was necessary to show how interwar intelligence had been too skeletal, and unfit for purpose. History should not be allowed to repeat itself. Cline's interpretation of history was a warning to those who might wish to suggest that dollars spent on the CIA might be better spent elsewhere.

It does matter that CIA partisans like Cline got carried away when discussing the past. That they believed so strongly in the OSS model is significant, as it was the inspiration for certain actions. As they matured in the embrace of their employer, the CIA, OSS veterans looked back on their youthful wartime service as a time when anything seemed possible and justified in the fight against Hitler's fascism in the context of all-out war. A significant number of them remembered the can-do approach of the OSS and favored direct action as a means of fighting the Cold War.

The OSS and especially the attitudes engendered *were* important to the creation and subsequent history of the CIA. If, however, we are looking for a full explanation of the agency's creation, it is helpful to remember that the CIA was an *American* institution. In furtherance of this approach, we can look more critically at the OSS and reconsider the interwar years that preceded its foundation We can see that the United States did, as Cline said, dismantle much of its World War I intelligence apparatus. But this was not a uniquely American demobilization.

Furthermore, while acknowledging the fall and subsequent rise of intelligence resources, it is vital to highlight some qualitative changes that the Cline school of historians ignored. One was the Department of State's disenchantment with espionage, which led to its abandonment of that practice. This left only the FBI as a civilian contender for intelligence leadership. FBI director J. Edgar Hoover made his bid to become a central intelligence tsar, but he lost out to the OSS, a pseudo-military organization that might not have been justified in peacetime and was acceptable only when war broke out.

There was certainly a downward swing in opinion concerning secret intelligence in the years following the armistice of 1918. It was partly the result of a desire to economize. War had put a strain on nations' economies, and there was widespread demilitarization. There was also a prevailing belief that when the shooting stopped in Europe, universal peace would break out. There was hope that the newly formed League of Nations would be a force for peace, even if the United States had refused to join it.

In 1921–2 Washington hosted naval reduction talks, an effort to curtail the arms race. Then, in 1928, the United States embarked on the Kellogg–Briand Pact. Named for Secretary of State Frank B. Kellogg and French foreign minister Aristide Briand, this was an initiative whereby a large number of nations agreed not to declare war against one another. Equally idealistic—or naive—were the Neutrality Acts that Congress passed between 1935 and 1937. A reaction to disclosures about the profits made by arms manufacturers in the last war, as well as to the argument that Germany had been unfairly portrayed as the sole aggressor in 1914, the laws cut off the supply of military and strategic goods to warring nations. The Neutrality Acts meant that the United States would be unable to stiffen European measures that might have stopped Hitler in his tracks.

One facet of interwar optimism was the view that spying did not belong in peacetime. In the UK, Sir Alfred Ewing, a founding father of wartime cryptography, declared that his team's espionage efforts had been "extemporized... as a war measure, to meet the need of the moment, ceasing to exist when its temporary purpose had been served."[5]

In the United States, faith in the morality of and necessity for espionage similarly plunged in the interwar years. As late as December 1938, a *New York Times* editorial expressed a commonly held, if soon to be dated view. Commenting on the Nazi spy case of that year that we shall discuss below, it asserted, "The spy problem is no problem at all." Increasing activities by America's counterspies were "symptomatic of the fear psychology which has swept the world rather than of the alarming importance of the international spy." The *newspaper* asserted that in the modern world there existed international transparency. That meant there were "few real secrets" anymore, so "The spy's role is limited." It warned against the militarization of espionage, which would be "alien to American traditions."[6]

The decline of American intelligence after World War I can be told as a tale of numbers. In 1918 the budget of the Military Intelligence Division (MID) was at its peak, $2.5 million. Personnel at MID headquarters numbered 1,441. By 1922, only ninety remained at HQ. The attrition continued until personnel numbers hit rock bottom in 1937, when only sixty-nine workers remained, getting by on a budget of

$267,000.[7] The Office of Naval Intelligence (ONI) continued to be active after the war and at a higher level than before 1914. It arranged for some of its officers to learn Japanese and kept a close watch on Tokyo's armed forces. Nevertheless, it too experienced major reductions from World War I levels. The backup staff at ONI headquarters declined from 306 in 1918 to 42 in 1920.[8]

U-1 suffered a succession of cuts. At the war's end the Bureau of Secret Intelligence ceased to exist. Some of its former personnel joined U-3. That subdivision of U-1 did survive various reorganizations and name changes to form the present-day counterintelligence and security arm of the Department of State. However, its attrition meant that State lost its capacity to send undercover agents into the field.

Then, in 1927, amid complaints that a "Harvard clique" was fixing jobs in the foreign service and that U-1 impeded the flow of intelligence from other sources, Secretary of State Kellogg announced, "The office designated U-1 is hereby abolished. All correspondence hitherto sent to that office will now be routed to U [the undersecretary of state's own office]." Alexander C. Kirk, who at the time was the head of U-1, had to write to the heads of military intelligence and to FBI director J. Edgar Hoover explaining the demise of his unit.[9]

In the wake of U-1's demise, the Department of State for a short time continued with its investment in another sphere of intelligence, the American Black Chamber. Then, in 1929, Secretary of State Henry Stimson withdrew his department's financial support. The Army had already been reducing its dollar input and building up instead its own signals intelligence capacity, and the American Black Chamber had to close down.

In contrast to the closure of U-1, the abolition of the Black Chamber stirred up long-term controversy. There were some persuasive reasons for Stimson's decision. They included cost-cutting and the telegraph companies' reluctance to infringe clients' confidentiality by handing over their messages to a government agency, which meant that the codebreakers had only a decreasing trickle of messages to decode. At the same time, the State Department was embarking on a long-term policy of distancing itself from clandestine activities that might have tarred its reputation for open dealing and trustworthiness. Stimson was at the time a strong advocate of trust and stated that the Black Chamber's eavesdropping was "unethical." One of his observations left a lasting impression: "Gentlemen do not read each other's mail."

It appears that the secretary of state was referring to American relations with friends in the white, "Anglo-Saxon" world, and not to nations like Japan. Moreover, he would make extensive use of signals intelligence when secretary of war, 1940–5. But his celebrated statement has added to the impression that American diplomats and indeed the nation as a whole were naive about spying in the interwar years.[10]

U-1 had been invisible in the first place, so its disappearance did not figure as a conspicuous event. It did leave a legacy in two ways. First, it was a precedent for centralized, civilian intelligence. Second, its surviving actors had some influence in future years. Just before U-1's dissolution, Leland Harrison stepped down as assistant secretary of state to take up ambassadorial posts. From 1937 to 1947 he was American ambassador to Switzerland, a hive of clandestine plotting in World War II. After dissolving his own unit, Alexander Kirk also progressed to a diplomatic career tinged with intrigue. When America's acting ambassador to Berlin in 1939, he cultivated the underground resistance to Hitler.

U-1's Ned Bell, the World War I US-UK cryptographic go-between, died of a heart attack in 1924. His daughter Evangeline, who spoke twelve languages and was no doubt attuned to the nature of her father's work, served in US intelligence in World War II. She romanced in London with OSS station chief David Bruce and married the man who would remain an influential figure after 1945.[11]

No doubt because of lingering memories of U-1, with the approach of World War II there were expectations that Adolf Berle would coordinate intelligence anew. He held the office of assistant secretary of state, a post that was the equivalent of the former office of undersecretary, in former days the home of U-1. Two weeks after Hitler's invasion of Poland, which prompted Britain's declaration of war on Germany on 3 September 1939, Berle received a visit from Sir William Wiseman, who had been the intelligence link between the US and UK in the previous war. Berle, who believed that the British had tricked America into entering World War I in 1917, rebuffed the veteran spymaster.

Berle and his fellow diplomats knew that trust was important to diplomacy, and they shied away from espionage, as any such association might make them seem untrustworthy. The idea of a diplomatically led intelligence community never quite faded away but was doomed to redundancy. In 1991, New York's Senator Daniel P. Moynihan proposed that the CIA be abolished, with the State Department taking over its intelligence role. Nothing came of this, which meant that State Department control of intelligence effectively ended in 1927 with the dissolution of U-1.[12]

Disdain for espionage was not limited to the State Department. It was widespread. Beginning in 1936, the Senate Committee on Education and Labor's Subcommittee Investigating Violations of Free Speech and the Rights of Labor (known as the La Follette Committee after its chairman) held a series of hearings that contributed to American hostility to the idea of espionage. Its 2.5 million words of testimony revealed that labor espionage was rife. In spite of previous denials, the Pinkerton National Detective Agency had on behalf of employers placed fifty-two spies in one union alone—the United Automobile Workers. The La Follette Committee showed that spying on workers conducted by employers and by agencies working on their behalf was endemic. The investigation was a reminder of the danger of spying on Americans at home—a lesson that was carried forward when the CIA was first barred from operating domestically, and then excoriated when it did.[13]

With the La Follette inquiry in full swing and enthusiasm for neutrality at its height, there now occurred an event that changed opinion about spies. In 1938, the FBI uncovered a Nazi spy ring. In spite of his professions of admiration and friendship, Adolf Hitler had authorized the subversion of the United States through anti-Semitic, pro-Nazi propaganda and through the theft of blueprints for new generations of US aircraft carriers, destroyers, and fighter planes. His agents sought and to some extent obtained the confidential defense plans for America's East Coast and for the Panama Canal. They compiled a list of prominent Jewish Americans, for example, New York City's Mayor Fiorello La Guardia, who would be dealt with in due course. The trial of four Nazi spies in the fall revealed a plot to open a "salon" in Washington, D.C., no expense spared, for the purpose of honeytrapping and blackmail.[14]

On the one hand, the revelations gave Americans yet another reason to hate spies. On the other hand, there was now an irrefutable case for counterespionage. The reputation of Germany had plummeted. The advocates of neutrality found themselves on the back foot. There was talk of preparedness. Very soon, an expansion in the size and role of the FBI was on the agenda. President Franklin D. Roosevelt still had to be cautious in light of the strength of the anti-interventionists in Congress, but soon after the conclusion of the spy trial he confided in Hoover that, as an emergency measure, his FBI's budget would be increased by almost 10 percent.[15]

There was a real prospect that the FBI, a civilian agency, as was appropriate, would develop into a central intelligence institution. In May 1940, Hoover called a meeting in his office to discuss intelligence cooperation. Undersecretary of State Berle recorded the event in his diary: "We had a pleasant time coordinating, though I don't see what the State Department has got to do with it." With that reservation, he was by now a supporter of the principle of mutual aid, and a few days later he claimed the credit for having converted the FBI and Military Intelligence Division to the idea of creating "a secret intelligence service" of the type that "every great foreign office in the world has."[16]

On 5 June Hoover accordingly proposed that the Military Intelligence Division, the Office of Naval Intelligence, and the FBI should coordinate their efforts, with the FBI in charge. The recently installed attorney general, Robert H. Jackson, supported the plan.[17] Soon afterwards, President Roosevelt approved the establishment of the FBI's Secret Intelligence Service. In 1936 President Roosevelt had already given Hoover's agency responsibility for coordinating intelligence on Latin America. From its establishment on 1 July 1940, the FBI's new secret intelligence service would be active in Central and South America in opposing Germany's foreign intelligence agency, the Abwehr, and suppressing indigenous left-wing movements. At the peak of its enterprise, the FBI ran 360 agents in the region.[18] By the summer of 1940, the FBI seemed to enjoy high standing in the White House and was well positioned to become the nation's first central intelligence agency.

It was at this time that Americans concerned with national security began to wonder whether they could learn from the British. The UK had a long—and much-vaunted—tradition of spying that was once more flickering into life. After World War I, there had been financial and personnel reductions that affected the domestic agency, MI5, and its foreign intelligence counterpart, MI6. Partly for this reason, British analysts received poor information about German capabilities. It gave rise first to underestimates and then to overestimates of Germany's military strength. By the years 1938–9, however, British intelligence and the quality of its estimates were beginning to recover, and by 1940 a year's warfare had further sharpened British intelligence capacities.

President Roosevelt had since the start of the war been conducting a secret correspondence with Winston Churchill and was aware of British know-how in the sphere of intelligence.[19]

The failure of Hoover's men to appreciate these developments spelled the end of his effort to make the FBI America's first central intelligence agency. In December 1940, senior Bureau official Edward A. Tamm sent his boss a misconceived memorandum. He reported that the president was "less and less sympathetic towards the British in the present war" but was keeping his opinions secret. He would "endeavor to manipulate matters to his own satisfaction without disclosing his viewpoint."[20]

Then, in January 1941, Hoover received the report of the Clegg and Hince mission. The mission's purpose, approved by Berle as well as Hoover, was to investigate British intelligence. In their report, Hugh Clegg and Lawrence Hince played to the Hoover agenda. They intimated to the expansionist director's receptive ears a welcome lesson—that British intelligence was now well financed. Other good news was that the British were a bulwark against communism and had no qualms about wiretaps and other forms of surveillance. But Hoover also passed on to President Roosevelt a further finding: Clegg and Hince said that MI5 had no control over local police forces and should be combined with MI6 in the interest of greater effectiveness. In other words, the British had nothing to teach the FBI about coordination.[21]

This proved to be the FBI's—and civilian intelligence's—last stand, at least for the time being. The president had by this time turned to a new kid on the block. Searching for an emissary who would report from Britain and come up with conclusions that differed from those of his defeatist, pro-Hitler ambassador to London Joseph P. Kennedy, Roosevelt fixed on William J. "Wild Bill" Donovan. Donovan was a lawyer and World War I combat veteran who, in spite of his Irish-American background, was a Republican, which suited FDR's search for national unity and bipartisanship. Another Republican, Secretary of the Navy Frank Knox, had recommended Donovan, as had William S. Stephenson, code-named "Intrepid," the Canadian head of Western Hemisphere British intelligence, who operated from a New York office and shared with Donovan a heroic status, having been a flying ace in the last war.

Appointed in June 1940, Donovan visited the UK and returned with the verdict that Roosevelt wanted to hear—Britain would be able to withstand any German military onslaught and was therefore worth supporting. In December, Donovan made a further trip to the UK. Ostensibly his goal was to secure information about the Eastern Mediterranean and North Africa, but he also made it his business to study British intelligence. He returned to the United States with the message that America needed a central intelligence agency.

In April 1941, Donovan proposed such an agency to Secretary Knox, and on 10 June he sent a memorandum on the subject to the president. On 18 June, Roosevelt issued an executive order establishing the Office of the Coordinator of Information (COI), and he placed Donovan in charge. Four days later, Germany launched Operation Barbarossa, its secretly planned attack on the Soviet Union that would spell the doom of Hitler's ambitions. COI was in no position to predict that. Then on 7 December Japan attacked the US Navy at Pearl Harbor. COI was unable to predict that surprise either, partly because turf-proud heads of other intelligence agencies were unwilling to share clues to the impending event with Donovan the upstart. The conclusion reached then, as repeatedly in the case of the CIA, was that central intelligence needed to be boosted.

The outcome was the Office of Strategic Services (OSS), established by executive order on 11 July 1942. The OSS was a military-style agency. At the same time, it was a reflection of American politics. President Roosevelt had established a number of agencies to tackle the Great Depression of the 1930s. Now, he had added another to tackle the crisis of World War II.

True to the traditions of U-1 and the Inquiry established by President Wilson, the OSS had a Research and Analysis (R&A) division. In charge was William H. Langer, a Harvard University history professor who would go on to perform a similar function in the early CIA. Like the Inquiry, R&A was a mini-university, drawing on academic talent across a range of disciplines. There were forty historians on the R&A payroll. Some of them would become distinguished after the war, seven of them becoming presidents of the American Historical Association—their presence in R&A helps to explain why there has been such emphasis on the role of the OSS as a precursor of the CIA. What these scholars did to help the war effort is debatable. They tended to duplicate the work of the State Department, and their influence on policy was minor. One historian concluded, "Although its intended clients were Secretaries and Generals, the ultimate significance of

R&A lies less in the fields of military or diplomatic history than in intellectual history."[22]

Some scholars treated their OSS employment as a kind of leave of absence that allowed them to write books that might have been significant but had nothing to do with the war effort. Arthur M. Schlesinger, Jr., for example, wrote a fine political study of President Andrew Jackson. Internal reports by Langer and others tended to measure their success by the numbers of personnel employed, and by the degree to which they had grown at the expense of their military intelligence counterparts. Nevertheless, R&A was clearly a predecessor of the Office of National Estimates, which would be an influential component of the CIA, and which made notable contributions to national security.[23]

More sensational and entertaining, once the veil of wartime secrecy had been lifted, were the tales of OSS's covert actions. These were extensive and varied. The OSS engaged in propaganda and subversion. Its agents trained with British Special Operations Executive (SOE) personnel in Camp X. The brainwave of "Intrepid," this was a training facility on the Canadian shore of Lake Ontario.[24] They also learned their craft closer to the action, in locations such as Arisaig, in the Scottish Highlands. OSS personnel became experts on everything from forgery to exploding donkey turds. Toward the end of the war, Donovan trained twenty-two assassins with the mission of killing German top brass like Adolf Hitler and Hermann Göring. Just as he was about to send a top-secret memorandum to General Dwight D. Eisenhower requesting clearance, he had second thoughts. He decided it would be better to kidnap his targets and bring them to trial. "Wholesale assassination," he explained to his colleagues, "would invite only trouble for the OSS."[25]

The OSS helped to shape the careers of four future directors of the CIA—Allen Dulles, Richard Helms, William Colby, and William Casey. A host of lower-ranking officers of the CIA owed their character development and attitudes to their experiences with the wartime agency. Captain Conrad "Connie" LaGueux is one example.

LaGueux commanded an OSS unit in the European war theater. In the summer of 1944, the Americans and their allies invaded German-held southern France, thus effecting a pincer movement to coincide with the D-Day landings in Normandy on 6 June. The OSS and the SOE were cooperating with the French Resistance, with the objective of creating mayhem behind enemy lines. On 6 August, an OSS unit code-named PAT parachuted into Virgule, a Resistance drop zone near the community of Vabre, in the department of Tarn. Operating behind enemy lines in conjunction with the local Resistance led by the Maquis de Vabre, PAT helped secure the surrender of thousands of German troops and to liberate the city of Castres.

LaGueux, the 22-year-old commander of PAT, was the Rhode Island-born American son of French Canadians. He spoke French fluently, as did several of his men. Barring certain romantic complications with local women, PAT got on well with the Resistance. The OSS had clearly selected the right unit for the job. LaGueux was decorated for his bravery and would in due course serve in the CIA's Far Eastern Division. Sliding from glory to humiliation, he was one of the last Americans to be evacuated from Saigon in 1975.[26]

The history of the OSS foreshadowed future characteristics and problems of the CIA. For example, the OSS had a competitive relationship with its British counterparts, and it tended to adopt indigenous peoples for its own purposes and then drop them. Both tendencies are in evidence in its effort to keep open the 700-mile trail from the Irrawaddy River in Burma to Kunming in China, intended to be a means of supplying China's resistance to Japan's invading forces. OSS's Detachment 101 recruited the help of the independence-seeking Kachin hill people. The object was not only to keep open the "Burma Road" but also to rebuff British attempts at local influence through another hill group, the Karen. In later years, the CIA continued to make promises to the Kachin to encourage them to fight Burma's communists, but they were ultimately abandoned to racial persecution by the Burmese (later Myanmar) government.[27]

In keeping with the appellation "world" war, the OSS operated globally. In Bulgaria, it unsuccessfully conspired with the businessman Angel Kouyoumdjisky to persuade King Boris III to reject the Nazis and switch to supporting the Allies in time to forestall an invasion and takeover by the Red Army. In Scandinavia, the OSS deployed its Norwegian Special Operations Group to attack bridges and other strategic targets, hampering the redeployment to southern war zones of Germany's occupying troops. One such raid flew in from Scotland and on the morning of 15 April 1945 blew up the Tangen bridge near Trondheim. Its leader was William Colby, the future director of the CIA. Such experiences were burned into the memories of the participants. Half a century later, Colby recalled the following proud conversation:

Q. Where were you on V-E Day, Granddad?
A. In the North Norwegian mountains...[28]

The OSS existed to find out the truth but also to distort it through "black" propaganda, both characteristics of the CIA. It experimented in regime change, and it excelled at self-promotion. Such practices also became behavioral characteristics of the CIA. The OSS, like the CIA, presented itself to indigenous peoples as the true champion of anti-colonialism, unlike France and Britain, while displaying nascent imperial characteristics of its own. The witticism "Oh So Social" played on the initials OSS and censured an Ivy League bias that had been evident in U-1 and would be conspicuous in the early CIA.

The bequeathing of personnel helped to ensure these OSS–CIA continuities. OSS alumni understandably took pride in their schooling. Lessons learned by the practitioners are not the same, however, as the conclusions a historian must draw. For whatever claims were made, the formation and character of the CIA did not stem exclusively or even predominantly from the OSS experience. The Secret Service, U-1, and FBI had all helped to shape an American civilian tradition that contributed to the shaping of the CIA. Yet the more military OSS did play an important role. It was double-pronged. Donovan's agency was a significant classroom for a good number of America's postwar spies. It also gave some of those spies false memories of infallibility, entitlement, and omnipotence that were out of keeping with the quieter, more thoughtful approach required of the modern intelligence analyst.

3

The Founding of the CIA, 1947

In 1947 there came into being the world's first democratically sanctioned secret service, the CIA. Gallup polls indicated a 77 percent public approval rating for the new venture. The democratic nature of the agency's founding gave it instant authority and standing.[1]

President Harry Truman might at first sight seem to be an unlikely person to have overseen the birth of a new intelligence agency. Traumatized by the death of President Roosevelt on 12 April 1945 and the sudden weight of high office, he famously urged a group of reporters who had assembled outside the Senate offices, "Boys, if you ever pray, pray for me now. I don't know if you fellas ever had a load of hay fall on you, but when they told me what happened yesterday, I felt like the moon, the stars, and all the planets had fallen on me."[2] Truman's three months' service as vice president under Roosevelt had not informed him about high office. Fond of holding all the strings of power, Roosevelt had kept him in the dark, and FDR's inadvertent successor knew little of state secrets and how to handle them.

Yet the new president was neither weak nor naive. He had faced enemy fire as a captain in the US Army in World War I. In the hardball politics of his native state, Missouri, he had run successfully for the US Senate and survived the scandal arising from his mainly innocent association with local political boss Tom Pendergast, who was convicted of tax fraud. He had learned the need to toughen up and that the world is not always a pretty place.

Though it took President Truman some time to adjust to the ever-changing challenges of the postwar world, he gradually developed a policy on peacetime intelligence. At its root were two principles. First, he did not want a permanent secret apparatus that would oppress the American people and deprive them of their liberties. Second, he did see the need to counter the challenge posed by the Soviet Union and thus the need for an enhancement of US national security. His support for a central intelligence agency arose from that security concern. By closely linking the CIA to the Soviet threat, he both increased its appeal at home and ensured that questions would be asked once the communist threat collapsed—as it did, four decades later.

Truman's first significant action regarding secret intelligence was to terminate the OSS on 20 September 1945. He admired neither the organization nor its director, William Donovan. There have been various explanations of his animosity toward "Wild Bill." One of them points to an erratic artillery bombardment on 14 October 1918. The barrage was meant to support the American attempt to capture the Kriemhild Fortress, part of the Germans' Hindenburg Line on the Western Front. As Major Donovan led his "Fighting Irish" troops into an assault on Kriemhild, Captain Truman's artillery unit misdirected what should have been supporting fire. There were American casualties. Shot in the knee in the course of the "Irish" advance, Donovan proceeded in peacetime to a lucrative law career and by the late 1920s was acting US attorney general. Truman's career advanced more slowly, and he could only watch with a feeling of worthless envy Donovan's heroic status and rocketing career.[3]

Other speculation centers on the story about FBI director J. Edgar Hoover spreading a malicious rumor about his career rival Bill Donovan. The rumor was that the OSS director was having an affair with Truman's daughter-in-law Mary. The truth is that while Donovan was close to Mary, there was no sexual relationship. His reputation for womanizing, nevertheless, gave wings to the story. The conspiracy theory holds that the rumor enraged the incoming chief executive. Whatever the reason, Truman did resent Donovan's pushy personality. However, he also based his OSS termination decision on the organization's less than distinguished wartime record. A fifty-nine-page report that detailed the OSS's failings arrived on his desk. The report had a military provenance and reflected turf-protective Army resentment of the OSS. But its very presence on the presidential desk suggests that Truman's objection was to the institution, not just its leader.[4]

Truman's initial reservations about a large-scale peacetime central intelligence agency were in tune with contemporary sentiment, as expressed by a vibrant segment of the press. The *Chicago Tribune*'s Washington correspondent, Walter Trohan, was to the fore. His views, disseminated through the *Tribune* and its siblings in the McCormick chain of newspapers, were critical of what was perceived as US foreign adventurism and of New Deal big government with its multitude of federal agencies. Trohan opposed the idea of a powerful central intelligence agency. On learning in February 1945 that Donovan was secretly advocating such an agency, Trohan had exposed the plan under the headline "New Deal Plans to Spy on World and Home Folks: Super Gestapo Agency under Consideration."[5]

The campaign by Trohan and like-minded journalists had support in Congress, and continued right up to the formation of the CIA, when the *Washington Times-Herald's* John O'Donnell denounced the arrival of a "super-duper gestapo-OSS-cloak-and-dagger organization."[6] At first sympathetic to such sentiment, Truman then moved away from it but could not escape its constraints. Reluctantly, he would come to accept that there would have to be democratic debate on, and sanction for, a peacetime secret intelligence agency.

This decision was unique. Other democratic nations had intelligence agencies, but they were shrouded in secrecy. The UK government did not even acknowledge the existence of its foreign intelligence agency, MI6, or its counterintelligence outfit, MI5. The democratic genesis of US secret intelligence was potentially a step forward for mankind. It was, however, thrust upon Truman. The president's initial preference had been to proceed without public consultation. To begin with, he did precisely that. Upon the dissolution of the OSS, he ordered a reduction of staff and Balkanization of resources. Research and development went to the Department of State, where Alfred McCormack presided over declining numbers; they included Herbert Marcuse, the Marxist whose works would inspire the 1960s New Left.

The State Department half-seized the opportunity. As it had on the eve of World War II, the department dithered over the prospect of housing a spying unit that might make it seem disreputable. It experimented with the idea of having secret agents in its embassies under diplomatic cover. These agents, recruited from an Army unit, were known as the Pond.[7] This may have been a derisive reference to La Piscine, the postwar French intelligence service nicknamed after a swimming pool near its Parisian headquarters, and it did not presage a good reception. Officially dubbed the Bureau of Intelligence and Research (INR), the Department of State's unit would, nevertheless, become a fixture in the intelligence community. Potentially, it mirrored

U-1, the coordinating unit in the office of the counselor that had operated between 1919 and 1927. But in 1945–6 State Department intelligence did not satisfy the ambitions of those who wanted a more powerful agency.

The OSS's secret intelligence and special operations branches went to the War Department to make up a new entity called the Strategic Services Unit (SSU). In charge of the SSU was General John Magruder, who had been the OSS's deputy director for intelligence. Some key ex-OSS personnel stuck with Magruder, for example, the future CIA luminaries Richard Helms and James Angleton. The SSU operated between 1 October 1945 and 19 October 1946, by which time Magruder had resigned because of the incessant reductions in budget—cuts that one would expect when a great war had ended but that were difficult to stomach. The SSU supplied a constant flow of intelligence reports about all parts of the world. By early 1946 it was compiling informed, if sometimes inconclusive reports on Soviet intentions. It is, however, unclear whether policymakers paid attention to SSU reports. Like other incumbents of the White House, Truman found he could not afford the time to pore over densely detailed reports with some-times no clear findings or recommendations. Early in 1946 he asked for a daily summary of intelligence, and there is little evidence that he paid attention even to that.[8]

To resolve these problems, the president fell back on a person with whom he felt comfortable and confident, someone he could trust. Sidney W. Souers was one of his home-state confederates, a group known as the "Missouri Gang." He was a graduate of Miami University in Oxford, Ohio. Though that picturesque institution was depicted as the Ivy League college of the Middle West, it was not as intimidating for the Midwesterner Truman as, say, private schools such as Yale or Princeton. A self-made refinance specialist, Souers had rescued the Missouri State Life Insurance Company from defaulting on its debts. In the war he had risen to be an admiral, but it had been as a reservist, and he was careful not to pull rank on the former Captain Truman. In the 1950s Truman recalled, "You can depend on this guy. He was one of my greatest assets."[9]

Souers had been designing a prototype for postwar intelligence for a while. He aimed for an agency that would be independent of the military, an ambitious goal because the Army and Navy comprised a powerful lobby with all the prestige of having just fought a victorious war. In January 1946, he mooted a scheme that would later be enshrined in law. A new intelligence institution would be responsible directly to the president, but be overseen by a National Intelligence Authority (NIA) on which the military was represented. Souers foresaw a need for covert operations. His agency would "perform such other functions and duties related to intelligence as NIA may from time to time direct." The Joint Chiefs of Staff were so enthusiastic about the proposal that it became known as the "JCS Plan."[10]

Three weeks later, Truman established the CIA's immediate precursor, the Central Intelligence Group (CIG). He devised an ambiguous pantomime scene for the inauguration of Souers as the first director. He gave Souers a wooden dagger, a black hat, and a black coat, calling him "Chief of the Gestapo" and "director of centralized snooping."[11]

Souers and his modestly sized staff of eighty tried to keep the president primed through the delivery of succinct summaries of tendencies and events that threatened national security. Almost immediately, they targeted the Soviet Union. Their fixation was a response to the times. In February 1946, in the immediate wake of the CIG's founding, Stalin announced that capitalism and communism were incompatible. In March, Winston Churchill selected Truman's home state, Missouri, as the venue for his speech announcing that an "iron curtain" had descended on Europe. In keeping with this outlook in both East and West, on 29 April Souers issued the CIG's first "tasking" directive: "There is an urgent need to develop the highest possible quality of intelligence on the USSR in the shortest possible time."[12]

The executive's concentration on the USSR continued right through to the formation of the CIA. In June 1946, General Hoyt S. Vandenberg took over as the CIG's director. He had the advantage of being the nephew of Senator Arthur H. Vandenberg of Michigan, a Republican who was key to the bipartisan support for the national security policies of the Democrat Truman. The CIG now underwent a rapid expansion. At first, General Vandenberg questioned the premise that the USSR was remorselessly expansionist. But by January 1947 his CIG was warning that Moscow's apparent restraint was merely a ruse to disguise its real intentions. Vandenberg helped prepare the way for the CIG's seamless transition into the CIA.[13]

All these discussions and arrangements took place behind closed doors. But in the early postwar years it was no longer possible to proceed without broader and more open support. In the recent war too many lives had been lost, too much suffering endured. GIs and their millions of family members, friends, and admirers expected a national security policy and that instruments of that policy would be accountable. They differed in this respect from British "Tommies," who had the reputation of having returned from the war tired of the Empire and eager for better social services.[14] America's veterans had not yet tasted bitter imperial fruits and were ready for the exercise of American power, so long as they were kept informed.

Truman realized that Americans with a wide range of different perspectives accepted the need for a central intelligence agency. Whereas President Abraham Lincoln had created the Secret Service by wartime fiat and President Theodore Roosevelt had authorized the FBI's precursor in a congressional adjournment, President Truman was able to preside over the creation of a central intelligence agency with the full blessing of Congress.

The president heeded public opinion in shaping the new agency. This helps to explain why he did not give a major role to J. Edgar Hoover and his FBI. Asked in a press conference why the CIG had no FBI component, he said, "We have to guard against a Gestapo. . . . You must always be careful to keep [national defense] under the control of officers who are elected by the people, then you won't have any trouble in the future." He reiterated the point in his 1955 memoir:"This country wanted no gestapo under any guise or for any reason."[15]

The atrocities of the Nazi regime refreshed a traditional American antipathy toward any spying arrangements that might result in a police state. Yet there were counterbalancing forces that ensured broad public support for espionage in support of national security. Ordinary citizens were less naive about espionage than one might suppose. There had always been public interest in spy movies and fiction, and World War II stimulated that interest. Notable espionage films of the era included *Confessions of a Nazi Spy* (1939), *Casablanca* (1942), *The House on 92nd Street* (1945), and *13 Rue Madeleine* (1946). In 1946 there appeared yet another edition of James Fenimore Cooper's novel *The Spy*, a story set in Revolutionary America.[16]

The popularity of non-fiction accounts of spying in America is further testimony to the sophistication of at least a portion of the nation's readers.[17] The rising generation of military leaders was aware of an American tradition of intelligence. Warner McCabe had lectured on that subject in the Army War College from 1940, and the Center of

Military History had housed a library of items on intelligence since 1885.[18] Over at the FBI, as we have seen, J. Edgar Hoover was a long-term advocate of central intelligence—even if, during the formative debate over the CIA, he found himself on the sidelines on account of the antipathies he had engendered in the snake pit of Washington politics. Similar, if complex memories existed within the Department of State, where U-1 was a precedent for INR.

In being civilian and centrally organized, U-1 was a precedent for the CIA as well. It is true that U-1 had been even more secret than the UK's MI6 and had not left an impression on the popular imagination. For insiders, though, some memories were vivid. For example, there was the case of Evangeline Bell. It will be recalled that Evangeline was the daughter of U-1's Ned Bell, who had in the wake of World War I pressed for a US codebreaking facility that would be independent of British guidance. Evangeline served in World War II intelligence and married David Bruce, station chief for the OSS in London and dir-ector of OSS operations in Europe. After the war, Mrs. Bruce was a member of the Georgetown set. This informal group included prom-inent figures such as James Angleton and George Kennan, who pressed for stronger intelligence. David Bruce issued a similar call while enjoy-ing a distinguished career as a diplomat.[19]

In spite of his rebuff at the hands of President Truman, Bill Donovan continued to press for his brand of a more powerful intelligence cap-ability. Together with several of his former wartime colleagues, he con-tinued to argue that something like the OSS was required. In public, he was circumspect. Addressing an assembly of journalists in February 1946, Donovan declared, "There is no moral justification for two nations at peace to break the other nations power of resistance or seek to sub-vert citizens into traitors. In time of peace, therefore, the true scope of intelligence is limited to the setting up of machinery that can gather, evaluate and interpret intelligence concerning other nations."[20]

Donovan's real ambitions were rather different. Urging on President Roosevelt the need for a peacetime central agency in November 1944, he had said there should be an ability to conduct "subversive oper-ations abroad."[21] In private, Donovan continued to argue for a covert action capability. In August 1947 he encountered a friend on the steps of Columbia University Library. This was Secretary of the Navy James Forrestal, who had just been appointed to be the first secretary of defense. There was an imminent need, he told Forrestal, for covert operations in France, where the communists were making advances. He was worried that the private anti-communist initiatives then under way would be chaotic: it would be "unwise to let this pass beyond your control."[22] In order to protect the world of free enterprise, Donovan saw a need to harness business to serve statist machinery.

Allen W. Dulles was another advocate of central intelligence whose views filtered through the morass of advice being given to the Truman administration. Dulles was a veteran of US intelligence in two world wars. He communicated his views to Chan Gurney, chair of the Senate Armed Services Committee, and Clare Hoffman, who presided over the House Committee on Expenditures. He wanted the forthcoming agency to be a vehicle for civilian intelligence: "It may well be more important to know the trend of Russian communism and the views of individual members of the Polit Bureau than it would be to have information as to the locations of particular Russian divisions."[23] Dulles had identified the holy grail of American intelligence. Knowledge of the Kremlin mindset would be a prime CIA objective, always sought, occasionally attained.

The foregoing factors help to explain the shape of the CIA and to a certain extent its formation. However, there was another essential cause. The inclusion of a provision for a central intelligence agency in the National Security Act of 1947 happened because of the support of Congress. That support had little to do with the Soviet threat, which nobody mentioned in congressional debates. It had much to do with memories of Pearl Harbor. Legislators were convinced that better US intelligence would have saved the nation from Pearl Harbor. They thought that it would be possible to build a new model that would save America from such shocks in the future. The truth is that studying past intelligence mistakes can produce models that are outmoded next time. Knowledge of the Zimmermann telegram did not prevent Pearl Harbor, knowledge of Pearl Harbor did not prevent 9/11, and knowledge of 9/11 did not prevent the Covid-19 catastrophe. In 1947, nevertheless, the nation's legislators were convinced that good intelligence would have averted Pearl Harbor and that proper reforms would prevent a repetition.

Congressional determination to avoid another Pearl Harbor is evident in hearings and floor debate that occurred in both houses in July 1947. Maryland's Senator Millard E. Tydings insisted on the need for central coordination—"otherwise, we may have another Pearl Harbor controversy." Representative Ralph Edwin Church of Illinois declared, "There is no better proof that we have been extremely backward in our intelligence work than the fact that we were completely surprised at Pearl Harbor." Such men remembered vividly the 2,403 lives lost on 7 December 1941 and hated the prospect of future military casualties. It is significant here that Senator Tydings was a veteran of the Meuse–Argonne offensive of World War I, and Congressman Church had two sons in the naval reserve.[24]

In the course of the democratic debate, there were plenty of objections. Two Republicans were to the fore. Senator Joseph R. McCarthy of Wisconsin saw the new agency as a military conspiracy and identified the CIA as his prime target in the years to come. Wyoming's Senator Edward V. Robertson also missed the point that the CIA would be civilian. He believed that the intention was to "create a vast military empire" with "all the potentialities of an American Gestapo." Rather more thoughtfully, Willmoore Kendall, who would become a notable neoconservative philosopher, was concerned about "the shadow of Pearl Harbor." He warned against a "compulsive preoccupation with *prediction*, with the elimination of 'surprises' from foreign affairs."[25]

Though it provoked varying reactions, Pearl Harbor was a prime cause of the establishment of the CIA. That rationale was significant, for it was not tied to a Soviet threat that would be of limited duration. The Pearl Harbor rationale, however irrational, conferred on the CIA the possibility of eternity.

Tydings, Church, and their congressional allies prevailed. The National Security Act of 1947 gave legal force to the CIA as part of a wide-ranging package of reforms. It made the Air Force a separate entity for the first time. It unified the Air Force, Navy, and Army under the new Department of Defense led by Forrestal. Under the terms of the act, the director of the newly created CIA would be responsible for the coordination of the entire intelligence community. A National Security Council (NSC) replaced the National Intelligence Authority that had overseen the CIG. Under the leadership, until 1950, of Truman's trusted stalwart Admiral Souers, the NSC supervised and tasked the CIA. The military had representation on the NSC and from time to time supplied the CIA's leadership. The CIA's first director, Admiral Roscoe H. Hillenkoetter (1947–50), had been wounded at Pearl Harbor, and the second, Walter Bedell Smith (1950–3), was a seasoned World War II general. But the principle remained that these military men wore civilian hats during their tenures. The CIA was independent and civilian in intent and spirit.

Gestapo neuroses had determined another of the CIA's essential features. Congress ordained that it would be no super agency and could operate only abroad. In a complementary move, the FBI was restricted to domestic work and, to Hoover's chagrin, had to surrender the intelligence jurisdiction over Latin America that it had enjoyed during and since World War II. The arrangement gave rise to counter-intelligence problems, as spy-catching demands domestic as well as foreign capacity, and the FBI did not always cooperate with the CIA. It would also mean trouble with Congress and the press when the CIA was caught stepping beyond its legislatively prescribed boundaries. Such transgressions would result in diminutions of the agency's standing in future years.

Once established, the CIA almost immediately broadened its activities. It became an agency that engaged spies and resorted to secretly influencing the affairs of foreign countries. This was nothing new in American history. President George Washington had spent a large part of the national budget on informational espionage, and President James Madison's secret agents had conspired to enlarge the territories of the United States at the expense of Spain. Recently, the OSS had made extensive use of "covert operations," and those memories in particular were fresh in the minds of those who molded the CIA.

Two factors help to explain why the CIA followed Bill Donovan's recommendation and resorted to "dirty tricks." The first was FDR's decision, in the wake of the Montevideo Conference of 1933, to renounce gunboat diplomacy. Overt military interventions in the internal affairs of Latin American states had cost the United States too much goodwill. Covert intervention was an alternative means of wielding power, so long as nobody found out.

The second factor was the intensification of the Cold War. The United States reacted to Soviet actions, for example, the denial of democracy to countries like Poland and the covert financial support of communist parties in France and Italy. American initiatives included the Truman Doctrine of March 1947, which gave military succor to Greece and then other countries, and, already in gestation in 1947, the Marshall Plan of March 1948, which afforded major economic assistance to Western European democracies, bolstering them against communism. The launching of covert operations or "psychological warfare" through the CIA was part of these US Cold War initiatives.

The operations came at an ultimate cost to the agency's standing. In 1963, remembering the humiliating failure two years earlier of the CIA's attempt to overthrow the Castro regime in Cuba by covert means, Harry Truman claimed he had never intended the CIA to engage in "cloak and dagger" operations, and he called for the CIA's remit to be narrowed to intelligence.[26] Such comments were a prelude to the 1970s theory that the CIA was an out-of-control rogue elephant. But in reality Truman had, along with other prime shapers of foreign policy like George Kennan, pressed the CIA into undertaking its clandestine activism.

No sooner was the CIA established on 18 September 1947 than it came under pressure to perform darker tasks. Secretary of Defense Forrestal asked the embarrassing question: did the CIA's remit extend to propaganda and sabotage? Section 102 (d) (4) of the National Security Act had instructed the CIA to "provide such other functions as the NSC might from time to time direct." Hillenkoetter turned to the CIA's counsel, Lawrence Houston, for legal guidance. Houston replied on 25 September that the contemplated activities "would be an unwarranted extension" of the agency's authority.[27]

However, the die had already been cast because of a legislative maneuver by an OSS veteran. Walter L. Pforzheimer was a Yale graduate, a bibliophile who took a keen interest in US intelligence history. In the war, he had laundered money destined for one OSS operation by passing it through the accounts of Yale University's Sterling Library. He had been counsel to the central intelligence group and advised on the drafting of the clauses in the 1947 National Security Act that established the CIA.

Pforzheimer withdrew a proposed clause that authorized "covert and unvouchered funds," as it would "open up a can of worms." He observed, "We could come up with the house-keeping provisions later on."[28] His furtive plotting meant that the CIA had not been entirely established in full view. Nevertheless, the agency had received the blessing of democratic approval. It was an unparalleled circumstance that gave it great authority.

4

Covert Action in the 1950s

The CIA immediately launched into a program of covert action that proved to be a gradually evolving disaster. It would alienate the majority of the world's nations and destroyed America's claim to moral leadership. At home, it threatened to dwarf the new agency's analytical functions. The covert action program in this way undermined the domestic standing of the CIA and its leaders' ability to promote, in US government, an intelligent understanding of world affairs.

But not quite yet. In the 1950s, the bomb fizzled but did not explode. Instead, the agency continued to enjoy the high standing conferred by its democratic birth.

The architects of the "dirty tricks" policy saw themselves as having good intentions that conformed with American tradition and the law. No less a figure than Thomas Jefferson had helped create the tradition. In the early years of his presidency, Jefferson sought to terminate the Barbary piracy that threatened American shipping off the coast of North Africa. To that end, he authorized and financed an attempt to achieve regime change in Tripoli, whose ruling pasha he had identified as a prime instigator of maritime crime. With the State Department washing its hands of the operation and ten US marines in charge, an informal army of 400 "insurgents" marched 500 miles across the Libyan Desert. Their approach had the desired effect in persuading the pasha to make peace overtures. As the historian Stephen Knott has shown, the Founding Fathers and their successors regularly launched covert enterprises of this type.[1]

In more recent times, the popular image of the OSS had been almost synonymous with covert operations. The OSS had undertaken such operations in parallel with America's enemies, friends, and companion institutions. No less a figure than commander in chief and future president General Dwight D. Eisenhower encouraged the Army to develop "psychological warfare." It was a term that the CIA would embrace.

A series of directives from the newly formed National Security Council launched America's peacetime program of covert action. The council's directive NSC 10/2, issued in June 1948, established within the CIA an office of special projects that would supply the bureaucracy systematically to administer dirty tricks. Rounding off a series of such directives in December 1955, NSC 5412/2 articulated the program's aims. It opened with the standard national security council mantra that it took "cognizance of the vicious activities of the USSR and Communist China." The CIA, it continued, would undertake "covert operations" that would create "troublesome problems for International Communism"; it would "discredit the prestige and ideology of International Communism"; it would combat the geographic spread of communism, and support "underground resistance" and "guerrilla operations" to undermine communist regimes, deploying "deception plans."[2]

The office of special projects and its successors launched covert actions. They also acted as a circuit-breaking mechanism. The plan was that the White House should be protected from any adverse fallouts. The circuit-breaking committee existed to take the blame when things went wrong. It was meant to enable the president to issue plausible denials of responsibility. President Truman even claimed, after the Bay of Pigs operation came unstuck in 1961, that he had never supported the principle of dirty tricks. That bit of sheer mendacity conformed to standard presidential protocols of denial. The truth is that, while in a very small number of cases the CIA may have acted without the say-so of the chief executive, Truman and later incumbents of the White House routinely authorized dirty tricks.[3]

Sometimes presidents supported covert action because it seemed like a quick fix in a potentially embarrassing crisis. However, in a fundamental miscalculation, presidents and their senior advisers also came to see covert action as a permanently useful and typical component of America's broad strategy. That strategy was to oppose the imperialism of totalitarian states, which in practice meant the Soviet Union and China. The strategy was integral to America's identity as a nation that owed its existence to the rebellion symbolized in the Boston Tea Party, an act of resistance to British imperial oppression. The strategy assumed an organic link between self-determination, democracy, and—a dogmatic but shaky assumption—capitalism.[4]

The CIA seemed to be in tune with American foreign policy. It was a means of enforcing the Monroe Doctrine, which in 1823 had stated that the "American continents... are henceforth not to be considered as subjects for future colonization by any European powers." Anti-Yanqui feeling had developed when the United States enforced that policy through gunboat diplomacy and interventions by the US marines. As part of his "good neighbor" policy, President Franklin D. Roosevelt tried to assuage such sentiment. And as we have seen, on the occasion of a pan-American conference in Montevideo, Uruguay, in 1933, he agreed to use no more military force.

At first Roosevelt's abstentionism came at no cost, for the United States had no reason to intervene. In the words of one historian, "The dictators who infested the area during his presidency never questioned Washington's hegemony."[5] When Nazi infiltration occurred, FDR changed his approach but still refrained from using observable force. Instead, he turned to clandestinity and sent in the FBI to combat both fascism and communism. After 1947, it would be the CIA's turn secretly to enforce Washington's will without—so it was hoped and intended—provoking an anti-Yanqui backlash.

The CIA operated within a framework of American principles expressed by recent presidents. President Woodrow Wilson had emphasized the principle of self-determination, but President William Taft, with his experience of having been governor of the Philippines, argued that new nations should become independent only after a period of education.[6] When President Franklin D. Roosevelt and Prime Minister Winston Churchill reaffirmed the principle of self-determination in the Atlantic Charter of August 1941, they did so with a similar reservation: former colonies would have to follow a "timetable" to complete independence, lest they make rash decisions and follow a leftward course. That reservation would entangle the CIA in many a postcolonial debacle.

Reaffirming the Atlantic Charter when the United States had entered World War II, Roosevelt and his international allies revived a further Wilsonian ideal, calling for a united nations organization. When the UN took concrete shape, it was very much an American concern.

Fifty nations agreed to its creation at a conference in Dumbarton Oaks, Washington, D.C., in 1944, and in the following year the founding convention met in San Francisco and determined that the UN headquarters should be in yet another US location, New York City, where it remains today. The intention was that America—and the CIA—would be in tune with the aspirations of an ever-increasing number of new nations determined to be fully independent of external control.

Harry Truman from the very beginning of his presidency made it clear that the United States would oppose the designs of the Soviet Union. The principles of his foreign policy were famously articulated by a serving diplomat. George Kennan was a member of the US ambassadorial mission in Moscow. He had first-hand experience of the stifling effect of state censorship. He played guitar and bass in an embassy orchestra known as the "Kremlin Krows" until the Kremlin objected, after which they became the "Purged Pigeons."[7]

A student of history, Kennan observed that the Soviet Union, in spite of its profession of a different ideology, followed a Russian foreign policy: "At bottom of Kremlin's neurotic view of world affairs is traditional and instinctive Russian sense of insecurity."[8] Ever fearful of being overrun by its neighbors and losing access to ocean trade routes, the communist leadership sought, in an aggressive manner, to surround itself with Moscow-aligned buffer states. Kennan recommended a policy of patient but firm "containment." This was less aggressive than the "rollback" policy favored by some, which would have meant a more militant effort at regime change in Russia's satellites, but it still envisaged an active role for the CIA. Like his fellow diplomats, Kennan did not want to sully the reputation of the State Department by association with spies, but he was a keen promoter of CIA covert operations.

On the eve of the CIA's formation, the president announced a policy that came to be known as the Truman Doctrine, giving concrete expression to the principle of containment. He was worried about the civil war then raging in Greece. In Greece, as in other Balkan states, the communists had played a leading role in resisting Nazi occupation, and those communists found themselves in control of the country when the Germans departed. Churchill had sought to reimpose the Hellenic monarchy by force. The civil war ensued. Soon after the Labour prime minister Clement Attlee succeeded the Conservative Churchill in office, the British government declared it had run out of money and could no longer afford to prevent the communists from seizing control of the Athens government.

Under the terms of his Doctrine, Truman stepped in with financial aid that ensured, in effect, that Greece would have a dictatorship of the right as opposed to a dictatorship of the left. He offered similar aid to Turkey and then to other nations. Two further initiatives confirmed and strengthened the policy of containment. They established a carrot-and-stick regime. The Marshall Plan of 1948 poured $12 billion into the economies of nations that followed the US prescriptive mode, and the North Atlantic Treaty Organization (NATO) was a military alliance that threatened wavering regimes as well as Moscow with the combined armed might of the new alliance.

America was by now the world's dominant power, and the history of the CIA became world history in the sense that the agency was a presence in just about every part of the globe. A comprehensive account of its activities would require several volumes, but a sense of their reach and purpose may be obtained by looking at some of their features.

The agency's activities were the responsibility of CIA "officers," US citizens who were salaried employees. The military term "officer" was a hangover from the days when Army and Navy officers, often in their role as military attachés, ran intelligence operations. It reflected the terminology used by the UK's MI(Military Intelligence)5 and 6, and by the OSS. The CIA was very much a civilian agency. However, the term "officer" still served to identify those in charge, as distinct from their "agents," who were mostly citizens of foreign nations.

One such officer was Frank Wisner, a rich and powerfully connected alumnus of the University of Virginia and the OSS, who directed US covert operations from 1948 until his nervous breakdown and death by suicide in 1965. He ran an octopoid enterprise with the aim of countering the propaganda of the communists' Cominform, established in 1947. Wisner filled a cultural void. For the United States had no equivalent of the Institut français (established in 1907) or British Council (1934), government bodies that openly projected favorable cultural images of France and Britain respectively. Congress was not disposed to fund such overt ventures, so under Wisner the CIA promoted cultural propaganda in secret.

Émigrés—those who had fled communist oppression in their own countries—were one of Wisner's assets. Or so he thought—for the émigré's perspective was often embittered and underappreciative of the social and economic injustices that had brought the communists into power in their native land in the first place. Another asset was the US communist apostate who "knew his Marx" and thus was supposed to know the mindset of the communist foe. An example here is Jay Lovestone, foreign policy adviser to George Meany, the American Federation of Labor's (AFL's) president. Meany would achieve a merger with the AFL's rival, the Congress of Industrial Organizations (CIO), in 1955 and serve as president of the AFL-CIO until 1979. Through the AFL-CIO's international activities, Wisner and Lovestone/Meany poured money into the coffers of anti-communist European labor unions.

The AFL-CIO was part of the American "voluntary" sector. The sector consisted of shell front organizations, but also organizations like AFL-CIO and charity societies that were otherwise free of government control, but which secretly helped Wisner and his colleagues achieve their aims. Most of the recipients of funds were not "witting," in that they did not know they were being subsidized. For example, some leaders of the National Student Association knew they were being secretly subsidized, but others did not. The association ran an international program, aiming to combat communist influence in student circles and organizations in Europe. In the UK, France, and elsewhere, students and others who attended drinks parties that stood out a mile in the still austere postwar decade would ask, "Who's paying for all this?" There was always a wag who would say the CIA, but nobody paid much attention. There was a conspiracy of silence in the name of liberty, or at least the next free drink.

There was scarcely any sector of society that the CIA did not exploit. Acutely aware that the communists had championed civil rights in the United States, the CIA supported a front organization called the Africa-America Institute. Leading journalists like Joseph Alsop worked with and for the CIA. In Alsop's case it was without pay. He mixed in the same Ivy League circles as the CIA leadership, and shared their exalted ideals and the determination to promote them. Estimates of the number of American journalists who worked with the CIA to shape a better image of America abroad and at home vary between 50 and 400.[9]

Turning its attention to women, the agency financed the activities of the Committee of Correspondence. This was an association of rich and influential American women who sought to combat claims that communism stood for peace, a cause dear to women. Encouraged by the CIA, the Committee publicized what it envisaged to be the fault lines in Soviet society. In April 1952 the Committee's first newsletter accused the Soviet government of forcing women to go out to work so that it could exercise "absolute control over the child with the opportunity to mold him into the pattern of well-disciplined little robots."[10]

In 1950, the novelist and former Comintern officer Arthur Koestler proved his worth as a fervent post-Marxist apostate. He organized a rally in West Berlin, democracy's enclave in the region that the Red Army had claimed at the conclusion of World War II and that now formed the short-lived state of East Germany. Out of Koestler's rally grew the CIA-supported Congress for Cultural Freedom. The first words in the Congress's "manifesto" were: "We hold it to be self-evident that intellectual freedom is one of the inalienable rights of man."[11] Slapping down its invisible gauntlet, the CIA decided that the headquarters of the Congress should be located in the left-leaning cultural capital of Europe, Paris.

The Congress had a spectacular agenda. It established offices in thirty-five countries. It covertly paid the production expenses of twenty magazines, including the then prestigious London-based *Encounter*, giving publishing opportunities to distinguished writers who thus addictively imbibed the laudanum of American-tinged freedom. It furtively disseminated propaganda through its own "news" service, complementing the activities of Radio Free Europe, another CIA initiative that aimed at Moscow's satellite states in Eastern and Central Europe.

The Congress for Cultural Freedom lured artists and musicians with paid-for exhibitions, performances, and prizes. In an "end of ideology" subterfuge, the goal was to weaken one ideology, communism, without admitting that this was with a view to promoting another, capitalism. European doomsters were horrified that the CIA's dollars appeared to have denatured Western culture, with representational art giving way to the meaningless daubs of abstract expressionism as practiced by that emperor with no clothes, the American painter Jackson Pollock. It was almost as if the CIA had taken on board the view of the 1920s African American cultural critic James Weldon Johnson that his white fellow countrymen had never produced anything "distinctively American."[12] In a crisis of confidence, so the argument might go, the Congress for

Cultural Freedom steered a nihilist course—better nothing than communism.

From another perspective, the Congress engaged not in nihilism but, as the senior CIA operative Tom Braden put it, in a "battle for Picasso's mind."[13] Surely, its argument ran, the main point of Pablo Picasso's artwork *Guernica* had been to attack totalitarianism, not to extol the left? The same sentiment dictated that George Orwell's novel *Nineteen Eighty-Four* should be packaged as an attack not on totalitarianism in the abstract but on communism specifically. The CIA translated and circulated copies of Orwell's works with this in mind.

A similar fate awaited *Doctor Zhivago*. Boris Pasternak's romantic novel of 1957 spanned the years between the Russian revolutions of 1905 and 1917 and could be interpreted as a humanistic attack on Soviet ideology. Moscow and the CIA both thought so. The former banned the novel, and the CIA promoted it. The Italian publisher Feltrinelli obtained a smuggled copy and published it in Italian, whereupon the CIA facilitated publication in the original language and found ways to insinuate copies into Russia. Already famous in Russia as a poet and translator, Pasternak became a household name in the West when he won a Nobel Prize. Then the David Lean-directed movie adaptation of *Doctor Zhivago* won several Oscars and remains to the present day a favorite Friday night weepie. The story of Pasternak's ascent to fame is even more interesting than the plot of his novel. The Soviets' claimed that the Nobel Prize was a propaganda fix and their instruction to Pasternak to decline the prize formed a script that the CIA could only have dreamed of writing.

It should be added that the agency took care not to cause injury to Pasternak. It was aware that "any obvious cold-war association of Pasternak with the West brought with it a risk of personal reprisal against him or his family." Private initiatives were to be discouraged and plausible deniability maintained:

> It was decided that exploitation of Dr. Zhivago would be undertaken solely by this Agency, because of the sensitivity of the operation, and that the hand of the United States government should not be shown in any manner.[14]

In all, the agency published around a thousand books.

Paris having been denatured by the dollar, the West's cultural capital moved to New York City. The wealthy governor of New York State, Nelson Rockefeller, had been a supporter of American intelligence for many years. He was both a financial supporter and a conduit for cultural Cold War funds. Now, his Museum of Modern Art became a go-to citadel of culture. Historians have debated the issue of whether, in the process, the CIA's cultural activities corrupted the arts and literature, with some of Europe's great intellectuals—like many of its politicians—being reduced to the status of capitalism's paid, cash-hungry fellow travelers. Or would Europe's scholars, in the end, have turned their backs on communism of their own accord, without any wastage of the almighty dollar? Of potentially greater concern back home, the CIA had clearly breached its charter's requirement that it should not operate domestically. But such debates did not surface in the 1950s and did not in that decade tarnish the agency's reputation and standing.[15]

The CIA was not just about propaganda. Its direct interventions in Italian and French politics in the late 1940s and the 1950s are well documented. It gave clandestine financial support to political parties, bought up ink supplies to stymie communist efforts to purchase printing presses, and generally tried to create a slightly tilted playing field in Western European politics. It is open to debate whether the CIA swung any election results. Hindsight suggests that, with the return of prosperity, European voters would have rejected communism anyway. The agency may have been wasting American taxpayers' money. Hugh Gaitskell, a leader of Britain's Labour Party, possibly swung to the right because of CIA money, but his real reason may well have been his perception that this was what UK citizens expected of him.[16]

In all three of the foregoing countries, the CIA invested in social democratic parties, a nod of support for Western European socialism. This "opening to the left" was in some degree opportunistic. Supporting the moderate left was a way of keeping out the Marxist dogmatists. However, conservative US politicians detected not pragmatism but a continuation of New Deal socialism conducted by a privileged Ivy League elite whose cosmopolitanism smacked of treason. Senator Joseph McCarthy identified the CIA as his number-one target in the national anti-left witch-hunt to which he gave his name in the early 1950s.

Allen Dulles, director of the CIA 1953–61, stood up to the intimidating populist. He refused to allow his officers to testify before McCarthy's committee investigating domestic communism, and Vice President Nixon had to step in to broker a deal. McCarthy retained the right to subpoena CIA witnesses, but only on the understanding that he never would.

The CIA may indeed have reflected the American leftist tradition in being "soft" on Europe's social democrats. But was it also soft on Europe as a region because Europeans were white and the cousins of so many Americans? Did that racial attitude apply also to other predominantly white parts of the globe like left-leaning Israel—and like Canada and Australia? The last two, along with the UK and New Zealand, were also members of the American-led "Five Eyes" club, the outcome of a 1946 agreement that gave members access to advanced signals intelligence.

When it came to toppling regimes, the CIA was indeed more circumspect in dealing with Europe than it was in the world's other regions. Its "Health Alteration"-cum-assassination Committee appeared not to be active in white nations. There was a more conservative approach even to the communist dictatorships behind the Iron Curtain. In October 1949 the CIA did act with British intelligence in an unsuccessful attempt to infiltrate anti-communist expatriates into Albania to overthrow First Secretary Enver Hoxha's regime. However, that operation was an aberration.

Less because of voluntary restraint than on account of entrenched resistance within the dictatorships and, from October 1949, Moscow's possession of a nuclear deterrent, what was routinely called a "crusade" against Eastern European communism more closely resembled remonstrative finger-wagging. In 1956, Hungary rose in rebellion against Soviet control. The uprising failed in one sense when the tanks rolled into Budapest. The tragedy became a triumph in another way, for the brutality of Moscow's repression discredited the communist cause worldwide. For this, however, the CIA could claim no credit. Lacking assets in Hungary, where it had only one agent in post, it had missed signals that peaceful reform might be possible and then advised against the uprising.[17]

In non-European, non-white parts of the globe, the CIA—guided, it must be remembered, by the White House—engaged in more ruthless behavior. One reason for this is that there was greater opportunity. Moscow, demonstrating its own brand of nativism in spite of communism's theoretical commitment to antiracism, was less likely to resort to nuclear retaliation to defend peoples in those parts of the world or to deploy conventional military force. Another reason for the employment of dirtier tricks in the "Third World" was that, in spite of the vibrancy of its civil rights movements, the United States had not fully divested itself of the prejudices of older imperial powers.

The CIA's Third World interventions were not always more radical than in Europe. For example, the Philippines were the scene of a relatively benign application of a covert operational variant, counterinsurgency. President William Taft (1909–13) had formulated a US policy whereby the islands would be administered as a colonial possession until such time as the Filipinos were deemed to be educated sufficiently to be trusted with democratic decision-making and autonomy.[18] That moment arrived in 1946, when the Philippines finally achieved independence. However, adherents of the Hukbalahap guerrilla rebellion did not respect the political formula bequeathed by the United States. The Huks had fought against US occupation for almost half a century and in the polarized days of the Cold War came to be seen as potentially aligned with the communists. To combat their insurgency, the CIA vested its faith in Edward Lansdale, a clandestine operations specialist who had served with the OSS in World War II.

Back in civilian life in California, Lansdale was a public relations man. During a resumption of his official clandestine duties in the Philippines between 1950 and 1953, he drew on his PR skills. Refining American counterinsurgency doctrine, he opposed the direct use of US military force—if you bomb people, he argued, they do not become your friends. Instead, weapons and military advice went to the indigenous people so that they could assume responsibility and take the fight to insurgents. Lansdale continued to favor education—in his time the US Army contributed to "psychological" warfare by building 4,000 prefabricated schoolhouses, a kind of mini-Marshall Plan that inculcated the right ideas and bought off potential Huk collaborators. Lansdale supported the politician that Washington favored, Ramon Magsaysay, the Philippines' defense secretary. By 1953, Magsaysay had all but defeated the Huks. In November of that year, he was elected president of the Philippines. Lansdale and the CIA had contributed to regime stabilization by respecting indigeneity. Such respect would be in short supply in other US foreign interventions, beginning with Iran.

The August 1953 Iranian coup took its place in CIA folklore. At its root was the issue of "nationalization," a word and concept that awkwardly chimed with nationalism, a useful tool against communism, but also with socialism. Iran's prime minister, Mohammed Mosaddeq, championed the policy. Wildly popular with his fellow Iranians, Mosaddeq was an impassioned advocate of anti corruption, transparency, democracy, and a limited monarchy. He was also opposed to foreign manipulation. "A nation that cannot administer its own house without the aid of others," he declared, "is unworthy of living."[19]

What "Orientalist" and uptight Westerners saw as self-indulgence and emotional instability—Mosaddeq conducted affairs of state from his bed and was given to fits of dramatic fainting and weeping—endeared him even more to citizen voters who could see that he put Iran first. That meant he upheld his nation's traditional opposition to Russian encroachment. He would have no truck with Soviet overtures, which made him popular at first with President Dwight D. Eisenhower. But the Tudeh, the Iranian communists, had pressed for Iran to insist on a share of the revenues received by the British Anglo-Iranian Oil Company, whose domination of Iran's oil resources was a legacy of empire. A sister oil company in Saudi Arabia had settled on a fifty–fifty share, but when dealing with Tehran Anglo-Iranian would not budge.

Mosaddeq decided to nationalize all of Anglo-Iranian's assets. In the overheated political climate of the McCarthyite era, his action could be interpreted as "communist." It also threatened more widespread economic consequences, for other nationalist leaders might be encouraged to take over assets belonging to Americans. It certainly put the Tehran leader on a collision course with Winston Churchill, who was back in office in the UK as Conservative prime minister. Churchill instructed MI6 to organize a coup. Eisenhower, having had a change of heart, gave the go-ahead for a joint venture, and the CIA took its instructions from Undersecretary of State Walter Bedell ("Beetle") Smith.[20]

Smith, having recently served as director of the CIA (1950–3), was well versed in clandestine affairs. MI6 and the CIA prepared the ground, rioters staged an insurrection with the loss of 200 lives, and that forced Mosaddeq out of office. The CIA's station chief in Tehran reported that there had been "a large element of spontaneity" in the rioting, but in truth his agency had paid many of the insurrectionists.[21] Muhammad Reza Pahlavi, a man who claimed the right to be the hereditary ruler of Iran, returned from exile and would govern as an absolute monarch with policies favorable to the United States. It was like a reverse movie of the American Revolution, with George III back in control. Although the CIA had engaged in the operation rather late in the day, Kim Roosevelt, its senior officer on the scene, claimed credit for the whole operation.

Iran was the first of two rapid-fire CIA involvements in the overthrow of democratically elected leaders in the "Third World." The next example occurred in Guatemala. There, 37-year-old Jacobo Árbenz had become the democratically elected president in 1951. He promised to end Guatemala's neocolonial dependence on the United States and to raise from poverty the historically oppressed peasantry of his nation. The Boston-based United Fruit Company had large land holdings in Guatemala and paid minimal taxes because of deals with previous, corrupt regimes. Árbenz nationalized United Fruit's property, with the intention of distributing the land to peasant families, and offered the company $1 million in compensation. This sum was based on the previous low valuation for tax purposes that United Fruit had extorted from previous regimes. Unabashed, United Fruit demanded $16 million and started a campaign of vilification in the United States. "El Pulpo" (the Octopus), as the firm was known in Latin America, was well connected in Washington. Undersecretary of State "Beetle" Smith was on its board of directors, and it drew on the law services of a firm presided over by Secretary of State John Foster Dulles and his brother Allen, director of the CIA.

When the Soviet dictator Joseph Stalin died in March 1953, the Guatemalan Congress held a memorial service. In vain did Árbenz's tactless wife Maria explain to the equally insensitive US ambassador John Peurifoy that there would also be a service for Winston Churchill when his time came. Jacobo Árbenz was not a member of his country's communist party but, as the historian Richard Immerman put it, he failed the "duck test." The "blind" anti-communist text ran as follows: "If he quacks like a duck and waddles like a duck, you just assume he's a duck."[22] Worse still, Árbenz's land reforms were a popular success. There was a danger that social democracy would succeed. And this was Latin America, not Europe. A CIA memorandum referred to "an intensely nationalistic program of progress colored by the touchy, anti-foreign inferiority complex of the 'banana republic'." The policy-makers allocated $2.7 million, not for an opening to the left but for the purpose of regime change.[23]

In November 1953, the CIA drew up its plan. As in all cases of this kind, it allocated a name to the scheme, in this case PBSUCCESS. The plan aimed to replace the "communist-controlled government of

Guatemala" with a "pro-US government of Guatemala."[24] In the event that propaganda and persuasion failed, the program would implement a policy of "executive action," or assassination. There were fifty-eight names on its "list of disposees," government officials and left-wing leaders who were invidious to the CIA's plotters. For the use of local assassins, the CIA prepared a manual called "A Study of Assassination." The educative leaflet considered the merits of hammers and wrenches as murder weapons and pointed to the advantage of "severing the spinal cord in the cervical region." President Eisenhower's CIA thus devised what may have been the first-ever US program of murder as an official instrument of foreign policy.[25] In the event, such drastic action proved to be unnecessary, as PBSUCCESS implemented another phase in its regime change plan: "para-military force enters target country, proclaims authority, declares target regime null and void."[26]

Once the Arbenz government had collapsed, the CIA installed the regime of Castillo Armas. The puppet dictator suspended the Guatemalan constitution, repealed Árbenz's land reform, and arrested and murdered his enemies. By the time Armas was himself assassinated in 1957, Guatemala had become a destabilized country. Outside the United States, the world was beginning to take notice of the new policy trend. France's leading newspaper, Le Monde, commented on "the hypocrisy with which some Americans condemn the colonialism and the subversive plots of others" while using similar methods to "get rid of governments they do not like."[27] Inside the United States, the CIA's secrets remained safe for the time being.

Operations that had the effect of destroying America's reputation internationally had a different effect in Washington's circle of power brokers. The CIA developed an air of invincibility. This endured even when attempts at regime change failed in the second half of the 1950s. Christian-centered, CIA-sponsored plots to overthrow majority-Muslim Syria's left-leaning, fragile democracy cost the US taxpayer $3 million. They failed in both 1956 and 1957, driving Damascus into the arms of the Soviet Union for succor. In Indonesia, the nation with world's largest Muslim population, the CIA sought to unseat President Sukarno in 1958 and thereafter. Its attempts to bomb his forces into submission and to smear him as a pornography addict having failed, the democratically elected Sukarno sought to remain in office by imposing "guided democracy," and by leaning ever more heavily on communist support, which led to a bloodbath in the 1960s.[28]

Then, in July 1960, at the instigation of colonialist Belgian plotters, the mineral-rich province of Katanga broke away from newly independent Congo. The 34-year-old prime minister Patrice Lumumba failed to secure UN support and appealed to the Soviets for aid to hold his country together. In response, the CIA's Richard Bissell authorized the delivery to Leopoldville of a syringe for the purpose of injecting a "lethal biological material" into Lumumba's toothpaste.[29] Although this particular attempt came to nothing, compliant local assassins lined Lumumba up against a tree and shot him. Debate still rages about the possible involvement of the Belgian colonial Sûreté, or security force.[30]

The impact of covert action on US foreign relations in the years up to 1960 invites consideration. According to one theory, Moscow hardened its policy toward its satellite buffer states because of the threat of Washington's Cold War tactics, crushing national aspirations and the prospects for free elections.[31] That may be so to a degree, although the trajectory of repression was well established before the arrival of the CIA.

More seriously, international opinion drifted away from Washington. From the first session of the UN General Assembly in 1946 to around 1960, the United States was able to control its own creation and remained committed to it.[32] The admission of newly independent Third World states changed the situation, for it was inescapably the case that US covert policy was particularly ruthless and objectionable in non-white regions of the world. In 1960, the UN acquired sixteen new states from Africa. This heralded problems in the General Assembly and, in due course, a loss of US enthusiasm for the great internationalist enterprise it could no longer control. The so-called "success" of the Iran and Guatemala ventures had contributed to this failure.

At home, in contrast, the Iran and Guatemala outcomes contributed to a high degree of confidence in the CIA. It was like remembering a game of hockey in terms of the goals one's team scored without regard to goals conceded. As for wider public opinion, the American people remained in blissful ignorance of the CIA's covert objectives and methods. The congressional subcommittees charged with oversight remained mute. Even if citizens had been better informed, they might have remained passive in the decade when McCarthyism, in its various shapes and forms, terrorized the nation. Donald James, a CIA officer who worked with the voluntary sector and helped to establish the agency's funding channels, recalled, "It was almost nobody that I couldn't go to in those days and say I'm from the CIA and I'd like to ask you about so-and-so and at the very least get a respectful reception and a discussion."[33] As of 1960, the CIA remained in good standing.

5

Intelligence in the 1950s

It was common to experience hassle on entering one of Europe's communist countries. Border guards would assiduously check for illegal currency importation, a reflection of the rigged economies of the totalitarian nations. Cameras were best kept hidden, for to display them risked forfeiture or even a ritualistic smashing of the artifact.

Travelers formed the impression that the regimes were behaving in a neurotic manner, and there was some truth in that. Behind the smashed camera, however, there was a rational assumption. Moscow and its satellites had reasonable grounds for suspicion. The CIA did ask traveling businessmen and tourists to photograph military installations, sometimes issuing, for example, instructions on when and at what angle to take photographs when flying over Soviet territory. It was one way of finding out what was happening in the closed and secretive regimes that lay behind the Iron Curtain. The agency was doing its *intelligence* job.

Not that there was uniform agreement on the precise nature of that job. It will be recalled that in Congress memories of Pearl Harbor had governed the debate, and speaker after speaker had demanded the avoidance of another surprise attack. The CIA would indeed avert surprises, though we do not know the whole story here, for an intelligence agency will try to make a secret of its successes and the methods by which they were achieved. The avoidance of every surprise was, however, impossible. Critics of the agency would list its failures with regret and sometimes glee. Examples included its non-prediction of the communists' blockade of West Berlin in 1948–9, of North Korea's attack on South Korea in 1950, of popular uprisings in communist East Berlin in 1953 and in Hungary in 1956, and, in the latter year, of the Anglo-French-Israeli surprise attack on Egypt, an attempt to repossess that profitable vestige of European imperialism, the Suez Canal, following its nationalization by the Abdel Nasser regime.[1]

Early on, when the fledgling agency gave no warning of the outbreak of rioting in Colombia's capital city in 1948, the intelligence veteran and philosopher Willmoore Kendall observed, "The shadow of Pearl Harbor is projected into the mists of Bogota." He warned against the "compulsive preoccupation with prediction, with the elimination of 'surprises' from foreign affairs."[2] The CIA took Kendall's advice. While it did apply itself to the anticipation of surprises whenever possible, it focused largely on what had agitated President Truman when he promoted intelligence reform, the Soviet threat to US national security.

In pursuit of this task, the CIA suffered from certain disadvantages. As we noted above, the agency faced a closed society. Finding out about military dispositions in faraway Omsk and Tomsk was hard, getting into the Kremlin's mindset well-nigh impossible. George Kennan adapted biblical language in his pessimistic lament four years after the CIA's creation. Regarding prognostications about the future intentions of the Russian government, Kennan wrote, "We see 'as through a glass, darkly.' "[3]

Conversely, America was a very open society. Security precautions were in place at the CIA's headquarters in Foggy Bottom in the nation's capital, as they were after the move in 1961 to a custom-built campus in Langley, Virginia. America's arms industry and the military were amongst other sectors of society that also took precautions. However, in marked contrast to the situation east of the Iron Curtain, enemy spies were free to roam in the country, and they did.

Another disadvantage affected the CIA's office of research and analysis. The analysts had increased in number since the pre-CIA days but were not well qualified. In 1948, a year when US–Soviet tensions were deepening, only twelve of the Soviet and East European branch's analysts spoke Russian, only one had a Ph.D., and six had never made it to college at all.[4]

Reforms in 1952 created an Office of National Estimates under Sherman Kent, a leading intelligence theorist who would work toward higher standards. Nevertheless, in 1954 President Eisenhower asked General James H. Doolittle to prepare a report on the CIA's efficiency, and the Air Force man concluded that there was too much "dead wood" at the agency, time-servers who were too inert to move on to other occupations. He noted that there was a particular problem in recruiting academics. Especially in the sciences, scholars' promotion depended on the constant production of articles, so those who were able and ambitious avoided putting in a spell at the CIA, when they would be nonproductive academically.[5]

The agency's 1950s scientists would come to be remembered more for their contributions to the development of psychedelic drugs and biological toxins than for their brilliant deductions about the potency of Moscow's arsenals.[6] To underestimate the CIA's technical capabilities would, however, be a mistake. The agency's Division D furnishes an illustration of the agency's technical potential. Created in 1947 to liaise with signals intelligence before and after the creation of the National Security Agency in 1952, Division D developed a life its own and supplied the CIA with the fruits of worldwide telephone tapping, as well as with other products of its own signals and communications intelligence efforts.[7]

The CIA's advantages, in fact, outweighed its weaknesses. One such advantage was that the United States was a nation of immigrants. Although America's millions of Russian and Polish speakers were potentially enemy assets, in the overwhelming majority of cases they were patriotic and could have supplied a rich stream of translators. The same could be said for Chinese Americans and other arrivals from Southeast Asia, and for Arab Americans. Whether for class and status reasons or because of populist prejudice, the CIA did not fully exploit its foreign-language speakers, who nevertheless remained a priceless national advantage.

The "rat line" at least partially offset the absence of sources inside the communist countries. Forgiven for having been Nazi "rats," German intelligence officials who had specialized on the Soviet Union were now useful allies. Gustav Hilger, for example, had been in Hitler's foreign office and was one of those who helped the CIA. The prime case, though, is that of Reinhard Gehlen. General Gehlen had run military intelligence on Germany's Eastern Front in World War II and had a deep knowledge of the Soviet foe. In 1946 US occupying forces in Germany established an anti-communist unit called the Org, with Gehlen in charge. The *Abwehr*, or German foreign intelligence, had spied on America until its dissolution in 1945. But America in the mid-1950s allowed West Germany to revive its spying capabilities under the guidance of its close ally, General Gehlen. With US encouragement,

Gehlen served as head of the German Federal Intelligence Service (Bundesnachrichtendienst, or BND) from 1956 to 1968.

"Walk-in" defectors from the communist bloc would further bolster the CIA's intelligence effort. Peter Deriabin was one such defector. In 1954 he concealed himself on a westbound freight train and rode the rails into Austria. Deriabin was a veteran of the NKVD and SMERSH police/intelligence organizations and thus a useful informant for the CIA when he decided to make a new life for himself in the United States.

As evidenced in the democratic nature of its creation, the CIA enjoyed public support. That support was important in allowing the agency to exist and operate, and because it ensured a continuing a continuing flow of congressional appropriations. It meant that private citizens would cooperate with the CIA if invited to do so. The CIA's popularity encouraged the White House to trust and listen to the agency's leaders, and to take its estimates seriously.

Potentially, more people on the political right might have joined in Senator McCarthy's attack on the CIA with its opening to the left. But even the *Chicago Tribune*, hitherto a reliable opponent of big spy agencies including the CIA, was going through a change in ownership that muted its message. In the case of the *New York Times*, its publisher Arthur H. Sulzberger was out of step with developing US diplomatic ties with Israel in being a Jewish anti-Zionist, but he was apparently still very much in tune with the CIA, as he reportedly allowed agents to pose as *Times* journalists.[8]

Hollywood was a potent image-shaper. In the aftermath of World War II, its studios addressed the subject of the OSS. The movies *OSS* and *Cloak and Dagger* both appeared in 1946 and romanticized the wartime precursor of the CIA. In 1947, the year of the CIA's creation, 20th Century Fox released *13 Rue Madeleine*, starring James Cagney. Here, the OSS played a heroic role in supporting the French Resistance. Such films reinforced the idea that the OSS was the parent of the CIA and the guarantor of its excellence. They shone a favorable light on covert operations, at least those undertaken in wartime.

Under pressure from McCarthyism, Hollywood then switched to domestically focused anti-communist spy movies that did not focus on the CIA with its international remit. In *Big Jim McLain* (1952), John Wayne played the part of an investigator for the House Un-American Activities Committee. In *My Son John* (also 1952), communists machine-gunned to death near the steps of the Lincoln Memorial one of their own, a homosexual spy who had infiltrated the federal government and was about to cooperate with the FBI.

Meantime, the CIA went backstage. The agency had its man in Paramount Studios. Instead of facilitating glory movies about American spies, he devoted his energies to improving America's international image. He cleansed the content of films that would be screened abroad, for example, by removing images of American drunks and disparaging remarks about Muslims. He and his Foggy Bottom controllers kept the CIA out of the movies—in conformity with the need for secrecy, and ensuring there would be no celluloid criticism.[9]

Abroad, criticism of the CIA did gain momentum. The Englishman Graham Greene was one disparaging writer. His 1955 novel *The Quiet American* poured scorn on the CIA's work in Vietnam. However, the Hollywood director Joseph Mankiewicz heavily sanitized the 1958 movie of the same title. At home, the agency remained shielded from criticism.

Recent literature has challenged the view that congressional sub-committees charged with scrutinizing the CIA simply looked the other way. Congress could express its disappointment in a painful way. Between 1953 and 1955, a time when it was realized that the CIA would not be able to "roll back" communism as some had hoped and expected, Congress reduced its budget from $500 million to $335 million per annum. Most of those dollars went on covert operations, which left around $37 million for intelligence-gathering and analysis. Nevertheless, except for Senator McCarthy's quickly discredited charge in 1953 that the CIA was at the heart of a New Deal "crimson crowd" that was taking over the country, congressional support for the agency remained strong.

As the 1950s wore on, the nation got over McCarthyism with its domestic neuroses, and Congress became ever more appreciative of the need to be watchful regarding the military threat posed by the Soviet Union. By 1957, the CIA's intelligence directorate was receiving $157 million per annum, more than four times as much as in 1955. To this sum can be added money spent elsewhere. For the CIA's director was, under the terms of the National Security Act, responsible for the coordination of the entire intelligence community. So when the CIA's Office of National Estimates went about its work, its estimates took into account the findings of military analysts, the FBI, and other sources. One of these sources was the National Security Agency, founded in 1952 to coordinate signals intelligence—communications interception and codebreaking—in the three branches of the armed forces. In 1957 the NSA had a budget of $450 million. Although members of the intelligence community tended to squabble in a jealous manner, that did not detract from the fact that the CIA's analysts benefited indirectly as well as directly from generous spending.[10]

Every dollar at the CIA's disposal was an expression of public support. Yet while taxpayers' support was important to the agency's well-being, high standing in the White House was also a prerequisite for its effective intelligence performance. Allen W. Dulles, director of the CIA from 1953 to 1961, achieved just that. Dulles epitomized the privileged, upper-crust leadership of the early CIA. Educated at Princeton, he was the grandson of John W. Foster, who as secretary of state in 1892–3 had established the use of military attachés for intelligence purposes. At the time when Dulles first entered the foreign service in 1916 as a third secretary at the US Vienna embassy, his uncle by marriage Robert Lansing was secretary of state. In Vienna the young Dulles may have encountered and must have been aware of the renowned Evidenzbureau chief Maximilian Ronge, an authority on intelligence who depicted spying as a possible means of achieving peace.[11]

When the United States declared war on Austria–Hungary in 1917, Dulles moved from Vienna to Bern, in Switzerland, where he engaged unsuccessfully in a covert attempt to engineer a separate peace with his former hosts. He was, therefore, a not insignificant outlier of the Department of State intelligence organism that came to be known two years later as U-1. After an interwar interlude in his family legal practice, Dulles found himself once again in Bern. As an OSS agent in neutral Switzerland, he cultivated the anti-Nazi German resistance. His efforts to achieve peace by dislodging Adolf Hitler from power did not work out but were still impressive.

Eisenhower's choice as director of the CIA had thus worked with two intelligence models, that of U-1 and that of the CIA. The president still hesitated to nominate him, as his selection as secretary of state was Allen's brother, John Foster Dulles. A family monopoly of swathes of foreign policymaking might strike some as egregious. But there were obvious advantages in terms of fraternal liaison and of security. The Allen-and-John arrangement solved, for the time being, the problem of the State Department's reluctance to be linked to spying.

Allen Dulles's outgoing personality and gift for public relations were further assets. He would give an annual dinner for about twenty-five guests, half of them newspaper journalists. He was on friendly terms not just with the *New York Times*'s Arthur Sulzberger but also with William Paley of Columbia Broadcasting System, Henry Luce of *Time* magazine, and even the *Chicago Tribune*'s Walter Trohan, whose quixotic tirades about the danger of an American "Gestapo" had been a feature of 1940s press reporting.[12]

Allen Dulles had the ear not just of his brother and the press but also of the president. Allen was the only person allowed to wear carpet slippers in the White House (a concession to his gout). From the beginning, for example on the occasion of the death of Stalin in 1953, he would be summoned to the presidential residence to give his assessments. In the words of one CIA officer, Allen Dulles had "the American flag flying at his back and the President behind him."[13]

Right up to its dissolution in 1991, the Soviet Union remained the CIA's main intelligence target. An early problem for the CIA was its non-prediction of the successful outcome of the project that Moscow code-named "First Lightning." This was the first Soviet atomic bomb test, conducted at the Semipalatinsk test site in the satellite state of Kazakhstan on 29 August 1949. The agency had given 1951 as the first possible date for this event, and the earlier detonation caught the United States diplomatically unprepared. In terms of predicting future surprises, the agency had fallen at the first major hurdle.

Recriminations followed. Willard Machle, the head of the CIA's Office of Scientific Intelligence, had the remit of being watchful for weapons of mass destruction—biological, chemical, and atomic. As Kazakhstan's radioactive dust settled, he rushed out a four-page memorandum, "Inability of OSI to Accomplish its Mission." He accused the military intelligence agencies of withholding scientific data on weapon systems.[14]

There was another narrative, one that explains the early explosion of the Soviet bomb and Machle's inability to predict it. It is the story of how Klaus Fuchs stole America's nuclear technology in the name of communism and, undetected by the FBI, OSS, and CIA, gave the Kremlin's scientists a head start. Fuchs was an anti-Hitler German scientist who joined the communist party, then fled Nazi persecution, advanced his studies at the University of Edinburgh, and joined the Manhattan atomic weapon development project in New York City.

One day, he visited the Lower East Side clutching a tennis ball. It was a secret signal to Moscow spy recruiter Harry Gold, who then became Fuchs's controller as he delivered to the Soviet scientific community the secrets of American and British research.

Did the Fuchs betrayal make a difference? It is all too easy to jump to the conclusion that it did, if you think that communism cannot produce effective science. That was not the view behind the Iron Curtain. It was a matter of pride for the Soviets to believe in the unassisted efforts of home-bred physicists like Vitaly Ginzburg and Andrei Sakharov.[15] Yet it does seem probable that Fuchs aided the Soviet A-bomb and later hydrogen bomb programs.

The Fuchs case was also significant in another way. After an American tip-off, MI5 organized the arrest of Fuchs on 2 February 1950. For by now the scientist had left America and was head of the theoretical physics division at the British Atomic Energy Research Establishment, deep in the Oxfordshire countryside. After the arrest, J. Edgar Hoover fumed at "the sly British" for not allowing access by his FBI interrogators.[16] His reaction was a precursor of a great deal of Anglo-American recrimination on security issues—recrimination that, together with some additional embarrassing spy cases, undermined the standing of intelligence agencies on both sides of the Atlantic.

There were many other Soviet agents at work in the United States, and, sensational journalism being what it is, some of the trials of those accused of espionage caused a greater furor than the more serious Fuchs case. The most prominent examples are the trials of Alger Hiss in 1949–50 and of Julius and Ethel Rosenberg in 1951. Cumulatively, the spy trials contributed to McCarthyism, or the Great Scare, as it has been more accurately labelled, as Senator McCarthy was not its sole progenitor. If only briefly, the CIA fell under suspicion, accused of left-wing bias.

The Fuchs case contributed to the Great Scare, but also, potentially, benefited the CIA. Fuchs's betrayals could be seen as an exoneration of the CIA's non-prediction of the Soviet bomb test. However, that predictive failure was real enough and supported Kendall's view that the core value of the CIA did not reside in the elimination of all surprises. The agency would be better at other, longer-range types of forecasting.

The CIA strove to obtain information on Soviet military matters and intentions. To that end, it built a tunnel. It was a reaction to new Soviet tactics. Because of the West's interception of its radio communications, the Soviet military had been making greater use of cable transmission, thus going back to pre-radio days, but with better encryption and tighter security against wiretapping. To counter this, the British had built a tunnel into the Soviet-occupied sector of Vienna in order to tap their communications. Now, the CIA and MI6 hatched a plan to build a similar tunnel from free West Berlin into the Russian-occupied district of the city. The top-secret project was ready by 1955, with the 300-meter tunnel terminating in a listening chamber located under military communication cables in the village of Altglienicke, just inside the Russian sector. Between 11 May and 21 April the following year, the agencies' high-tech listening equipment harvested extensive data that would not otherwise have been obtained.

According to the CIA's partly declassified account of the venture, the American and British eavesdroppers learned about the dispositions of Red Army units and bombers. They learned the names of thousands of Soviet armed forces officers. They discovered that there were tensions between Russian forces and their East German and Polish counterparts. They deciphered communications indicating "Implementation of the publicly announced intention to reduce the strength of Soviet Armed Forces."[17]

One day, Western technicians at the far end of the tunnel dropped their coffees and bolted. They had heard ominous noises coming from above. East German and Russian soldiers were at work, digging. The excavators soon made contact with the fabric of the tunnel, and the game was up. Moscow orchestrated a great drama around the event. In East Germany, the *Schweriner Volkszeitung* published the first photograph of the exposure. Twenty-two other Party-controlled newspapers published implausibly identical "eyewitness" reports of the discovery to signal the "outrage" of ordinary people.

In the United States, by contrast, journalists marveled at the ingenuity of the tunnel's engineers and celebrated the fact that the CIA had come away with intelligence booty. *The Boston Post's* editorial writer was astounded:

Frankly, we didn't know that American intelligence agents were so smart. In fact, we were beginning to think that what the Central Intelligence agency needed was a few lessons on the fundamentals of espionage from some defected Russian agent. But now we take it all back.[18]

In 1961, the agency did learn some fundamentals from a defector named Michael Goleniewski. Goleniewski asserted that he was descended from the Romanov dynasty. He was actually born in Poland. It was while serving in Polish counterintelligence in the 1950s that he spied for the Russians on his Polish compatriots and then decided to spy for the CIA on both Poland and Russia. It was Goleniewski who told the story of George Blake and the Berlin tunnel.

A member of MI6 since 1944, George Blake had agreed to spy as a double agent for the Soviets' KGB a few years later. The British arrested him, and he confessed. He received a forty-one-year prison sentence, said to have been one year for every agent he betrayed. In 1966 he escaped from Wormwood Scrubs prison and ran away to the Soviet Union.

Blake had known all about the tunnel since before construction even started, and he had kept his Soviet masters fully informed. Speculation about the impact of his treachery has been rife. Did the Soviet and East German authorities withhold some information from the cable traffic that they knew the CIA was reading? That does sound probable. Did they issue disinformation in order to deceive the listeners about the plans and assets of the communist world? Sergei Kondrashev, the KGB general in charge of the Soviet agency's German department at the time, issued a post-Cold War denial. He explained that in the case of the tunnel the dangers of disinformation outweighed the advantages. For if the CIA had deduced it was being deceived and realized its opponents had been informed of the tunnel's existence, the search for a leak would have led to the double agent. Because George Blake was too precious an asset to squander, it would have been unwise to send out disinformation.[19]

There were other puzzles too. Why did the Soviet leader Nikita Khrushchev authorize the "discovery" of the tunnel in 1956? One possibility is that he needed a display of toughness in order to reassure the Party faithful, especially as he had recently delivered his de-Stalinization speech, which seemed to promise détente with the West, a "soft" policy.

One historian has suggested that the Soviet leadership allowed the CIA an insight into the closed Iron Curtain puzzle box because they "wished to convey" to the Eisenhower administration the message that Moscow's intentions were peaceful.[20] Perhaps, but even in today's more relaxed climate there has emerged no explicit evidence of this intention. Besides, it does not matter how, exactly, the CIA found out about the Kremlin's benevolent intentions. For Allen Dulles, the tunnel was "one of the most valuable and daring projects ever undertaken."[21] It was vitally important for him to be reassured and to be able to say that the Soviet Union did not pose an imminent mortal threat to the United States.

The unmasking of the tunnel plot was a setback for American intelligence, but the CIA director had further tools at his disposal. One of these was as far removed from the James Bond image of spying as one could imagine. It was economics.

Prior to becoming head of the CIA, Dulles had chaired an inquiry into how to coordinate intelligence and analysis. Its report in 1949 recommended the creation of a unit that came to be known as the Office of Research Reports. In due course, there was agreement with the State Department that the CIA's new office would take care of all economic research and analysis pertaining to the Soviet Union, with State looking after other areas of the world. The Korean War of 1950–3, in which the United States fought under a UN mandate, rang alarm bells as the prospect of war with the Soviets began to look real. Max Millikan, the Research Office's director, increased economic analysis as a matter of urgency. By 1953, the Office of Research Reports employed 766 personnel.[22]

Millikan formulated his office's objectives. They included:

Estimating the magnitude of potential military threats to the United States and its allies by calculating the level of military operations a potential enemy's economy could sustain.

and:

Assisting in judging the intentions of a potential enemy by monitoring changes in its economic policies.[23]

The idea behind the CIA's unit was that it should estimate the capacity of the Soviet economy to support the production of bombers, missiles, and other artifacts of war. It was no easy task. Official Soviet budgets were expressed in rubles, a currency whose nominal value was rigged, which made it an unreliable measurement of economic strength and productivity. Expenditures were routinely hidden from view or massaged for propaganda purposes. The CIA's economists devised algorithms to resolve these problems.

The CIA's economic-driven analysis countered inflationary pressures brought to bear on the Congress and White House. The lobbying for greater expenditure came from weapons manufacturers who wanted bigger arms appropriations to swell their profits. Pressure emanated also from the armed services and their intelligence branches, whose commanding officers sought promotion and aggrandizement. The services competed with each other, so that the Air Force wanted more bombers, the Navy more ships, and the Army more tanks. The result was a tendency to exaggerate what the Soviet Union had in the various categories as a justification of greater US expenditures. Seasoned, five-star general that he was, President Eisenhower saw through all this. In his January 1961 farewell address, he would warn against the influence of what he called the "military-industrial complex."

In the mid-1950s, in the mistaken belief that there was a "bomber gap," with the Soviet Air Force outstripping its US counterpart, Congress appropriated large sums to increase the size of the nation's B-47 and B-52 bomber strike forces. However, the CIA warned that Moscow was holding back from undertaking ruinously expensive and archaic investment in bombers. Its leaders were aiming to "leapfrog" the process to build intercontinental missiles and to explore space-based technologies. To conceal its methodologies, the CIA kept the public in the dark about the latter development, so the Soviets' launch in October 1957 of Sputnik, the world's first space satellite, came as a shock. There ensued a "missile gap" panic that was fanned by the arms industry and senior military leaders. The Air Force calculated that the Soviets would have 1,100 intercontinental missiles in place by mid-1961. The CIA's calculations destroyed the myth. The agency reckoned the Soviet economy could not support the production of more than fifty missiles per annum. Even this was an exaggeration, as in the event the cumulative number deployed would amount to no more than fifty.[24]

Meantime, the CIA was deploying its own high technology. Its U-2 high-altitude spy plane overflew the Soviet Union in 1956 and con-tributed data for the CIA's lower bomber estimate. In 1960, with concern about Soviet missiles now running high, the U-2 overflew the Tyuratam test range, but with inconclusive results. President Eisenhower knew that the Soviets were developing antiaircraft ground-to-air missiles capable of reaching the U-2's 70,000 feet flying altitude, but decided that the need for conclusive evidence outweighed the risk of a shoot-down. The experienced pilot Gary Powers agreed to fly a U-2 right across the USSR in a south-to-north quest for the missing data. As he was performing this mission, one of the new-generation anti-aircraft missiles disabled his plane. Powers bailed out, and the U-2 crashed onto Soviet soil. To the embarrassment of the United States, the Kremlin displayed the captured pilot and wrecked plane for the propagandistic purpose of showing that the nefarious Americans were breaking international law by violating Soviet air space.

In reality, the overflight and shoot-down in no way detract from the peaceful motives behind the mission. By such means, the CIA enabled a more evidence-based US defense strategy. Allen Dulles and his boss in the White House were hard-bitten war veterans who were unafraid to show restraint in the interest of peace. They respected each other, and there was respect for the CIA in the 1950s. The agency's high standing did not shield it from all criticism. But it did ensure respect for its intelligence product on the most important issue facing mankind, the nuclear arms race.

6

Bay of Pigs, 1961

The Bay of Pigs "fiasco," as it is usually described, was an unlearned lesson and a comet with a long tail.[1] What triggered the event was Fidel Castro's seizure of power in Cuba. The United States had backed the regime of the reliably right-wing President Fulgencio Batista. This dictator had run Cuba not just as a compliant friend of the United States but also as a partner of US mobsters, who took a significant stake in the island's flourishing casino and brothel industries. When Castro overthrew the Batista regime in 1959, it was not a foregone conclusion that the victorious insurgent would be a Moscow-leaning communist. His execution of 600 Batista supporters was alarming, but did not in itself flash a red light. It was when the newly installed leader abolished the vice trade and announced land reforms affecting US interests that he failed the notorious duck test. Washington began to treat him as a communist, and that is what he became.

President Eisenhower decided that Castro would have to go and entrusted the CIA with finding a solution. Richard Bissell, a veteran of the 1954 coup that had ended Guatemalan democracy, took charge in his capacity as the newly installed deputy director in charge of covert operations. He turned to the agency's Health Alteration specialist, Sidney Gottlieb, asking him to activate his chemical warfare arsenal.

"Dr. Death" did not have a stellar track record. His experimental development of the psychedelic drug LSD had ended in tragedy when the army scientist Frank Olsen took the substance and under its influence leaped to his death from a New York hotel window. Gottlieb now planned actions that would expose Castro to ridicule, for example, by administering a depilatory powder that would cause the leader's beard to fall out, taking his macho image with it. More seriously, through

Gottlieb's agency the world's leading democracy consolidated the place of assassination as an instrument of state policy. The scientist began with a plan for the disposal of Patrice Lumumba. Other assassins got there first, and Gottlieb's doctored toothpaste never touched Lumumba's lips. Castro now took the Congo leader's place as the CIA's prime target.[2]

Castro was a marked man from the moment he became president of Cuba. Plans to kill him have passed into popular mythology, and the stories typically waver on the borderline between fact and fiction. Take the case of Castro's mistress Marita Lorenz, one of the first to be entrusted with his dispatch. Her father Heinrich had been a German U-boat and ocean liner commander. He was a person of special interest to the FBI when the bureau investigated the Nazi spy ring of 1938.

Marita's mother Alice was a former Broadway actress. The Nazis sent her to the Bergen-Belsen concentration camp on suspicion of spying on behalf of the United States. Marita accompanied her mother and remained in the concentration camp until liberated by the arrival of American and Allied troops. At this point, 7-year-old Marita was raped by an American GI. After those experiences, where would her loyalties lie, and how reliable could she possibly be? Marita remained devoted to her father, but—like her mother—she became (according to her own account) an American spy.

In 1975, Marita told a *New York Daily News* reporter that in 1960 the US mobster known as Frank Sturgis (real name Frank Fiorini) gave her two lethal pills to administer to Castro. She hid the pills in her face cream. According to her account, Sturgis explained to her that he had been working for the CIA for the last decade.

How reliable was this story? Marita was an attention-seeker—her daughter Mercedes said she was just a concentration camp girl who wanted to be loved. For example, Marita told another story about a meeting she had in Dallas two days before President Kennedy's assassination on 22 November 1963. She claimed that the assassin, Lee Harvey Oswald, was present, as were Jack Ruby—who would kill Oswald—and Sturgis, together with E. Howard Hunt. The last two, by the time she told her story, were doubly infamous, having been arrested for the Watergate burglary, which led to the cover-up that led to President Richard Nixon's resignation. If Marita wanted to become a celebrity

by latching onto persons of notoriety, she succeeded. For example, the
1999 movie *My Little Assassin* is based on her story.

A 1977 CIA in-house historical assessment of the agency's assassination policy questioned the CIA link. It suggested, with no convincing
proof, that the CIA's cooperation with the criminal underworld was
not strongly developed until after the Bay of Pigs fiasco. Two years
later, a specialist staff report for a House committee came to a similar
conclusion. Perhaps the two pills did not come from Dr. Gottlieb's
store. All accounts agreed, though, that Sturgis set up Marita to use the
pills. They assumed that Marita told Castro who had dispatched her
and why. Marita later explained the reason why she did not kill the
Cuban leader: she loved him.[3]

The CIA's self-investigation indicated that before the Bay of Pigs
the agency merely went along with the Mafia crime syndicate, becoming actively involved in contracting hit squads only after the Cuban
debacle:

It is possible that CIA simply found itself involved in providing additional
resources for independent operations that the syndicate had already under-
way...In a sense CIA may have been piggy-backing on the syndicate and in
addition to its material contribution was also supplying an aura of official
sanction.[4]

The Mob did have its own motive to eliminate Castro, as he had shut
down their lucrative brothels. But a shadow of complicity nevertheless
hung over the CIA, and the agency's pre-Bay of Pigs efforts, if low-key,
were part of a documented and unsuccessful continuum. An evidence-
based Senate assassination inquiry in the 1970s found just eight CIA
attempts to murder Castro. The distortion worked both ways. According
to Cuba's counterintelligence chief Fabian Escalante, there were
altogether 612 plots to exterminate Castro between 1959 and 1993.
What all these reports have in common is the finding that the CIA
made repeated efforts to kill Castro, and failed.[5]

In February/March 1961, it was becoming apparent that the Dr.
Death approach to regime change was unlikely to succeed. With this
in mind, JFK and his CIA pursued a different plan, which led to the
Bay of Pigs invasion by proxy that took place on 17–19 April. President
Eisenhower had authorized the planning stage of the scheme, whereby
the CIA would organize and support a force of Cuban exiles who
would invade their homeland, which would prompt a spontaneous

uprising against Castro. In the fall of 1960, he urged the newly elected Kennedy to proceed, and JFK received historical briefing papers that reminded him of Eisenhower's plan:

17 March 1960 – President authorized Agency undertake covert action program to replace Castro government with one more devoted to true interests of Cuban people and more acceptable to U.S. in such manner avoid appearance of U.S. intervention.[6]

Victim to its own hubris arising from its earlier promotion of regime changes in Iran and Guatemala, the CIA leadership embraced the Eisenhower project and hoped that the incoming president would give the green light.

John F. Kennedy was a promising proposition for the CIA's leadership, as he was an insecure politician who aspired to prove that he was a worthy and courageous leader. Bravery was, in fact, ingrained in his psychological makeup. In August 1943 he had commanded a crew that heroically rescued wounded comrades from the Pacific waves after the sinking of their patrol torpedo boat PT-109. Wartime heroism was a political asset, and JFK did not object when his father Joseph organized a publicity campaign portraying his son as the *sole* hero of the event. Then, in 1956, by now a Democratic US Senator, JFK published a book on senatorial bravery. *Profiles in Courage* was the result of political calculation—for example, it promoted JFK's bipartisan credentials by praising the senator known as "Mr. Republican," the former presidential aspirant Robert A. Taft. The well-written book won the Pulitzer Prize, and JFK saw no reason to reveal that his speechwriter Theodore Sorensen had authored the work that appeared under his name and burnished his reputation for valor and literacy.

Though Kennedy had a liberal reputation, he reappointed Allen Dulles, by now a conservative icon, as head of the CIA. It was one of a series of moves designed to reassure Americans that the youthful president was no weak-kneed liberal. The reappointment meant that a CIA director wedded to the ways of the 1950s would now advise a young president who wanted to show that, where bravery and patriotism were concerned, he was the complete package.

The CIA assembled and paid a force of 1,400 Cuban exiles and trained them in permissive Guatemala. Nervous that Uncle Sam's hand would be revealed, Kennedy insisted that their landing site should be on the side of the island that faced away from the United States. Bissell and his colleagues quickly settled on a series of beaches on the Zapata Peninsula, dotted along the Bay of Pigs on the southwest coast of Cuba. On 17 April 1961, Operation Zapata commenced, and the expatriates began to disembark on the sands of their native land. They met withering fire from the defending militia, the Cuban Revolutionary Armed Forces. The invaders ran out of ammunition when the Cuban Air Force bombed their supply ships, and on 20 April they surrendered. The CIA having lost control of events, President Kennedy had to admit to US complicity and face criticism at home and abroad.

JFK asked General Maxwell Taylor to conduct an inquiry into what had gone wrong. The CIA's inspector general Lyman Kirkpatrick conducted a further internal investigation. The resultant reports identified the CIA's errors. Its operators had not properly consulted the intelligence directorship of the agency and should have realized that there was popular support for Castro in Cuba and little prospect of a sympathetic uprising. Bissell should have opted for a different landing site near hills, to which the landing force could have retreated in the case of stiff resistance. The CIA had arranged for air support by B-26 aircraft resprayed in the colors of the Cuban Air Force, part of the ruse to achieve plausible deniability and persuade the Cubans that their pilots had joined in the insurrection. But because the B-26s were indistinguishable from the real enemy, they came under "friendly fire" from the invaders' supply ships.

Twenty years after the event, Jack Pfeiffer's official CIA history of the Bay of Pigs episode offered an "evaluation of the evaluations." It depicted the Kirkpatrick and Taylor reports as overcritical of the CIA and as having a tendency to whitewash President Kennedy. By the time the Pfeiffer history was declassified in 2016, JFK was being routinely faulted for not sending in more air cover at a juncture in the battle when it was needed. There would have been nothing more to lose, as the "hidden" US hand in Operation Zapata was already plain to see. Eisenhower had taken such a step to ensure the 1954 regime change in Guatemala. Kennedy was open to the charge that he had lacked courage at a time when it was needed.[7]

Pfeiffer, like the earlier investigators, refrained from asking whether the overthrow of Castro should have been attempted in the first place. The inquiries having failed to inquire, little changed. Even after Dulles and Kennedy had departed the scene, the CIA continued with its clandestine interventions in the Americas. Two of the CIA's interventions in 1964, the first full year of Lyndon B. Johnson's presidency, fell short of paramilitary action but were still ominous for the health of democracy in the Americas. In September, clandestine CIA spending helped to buy the presidential result in Chile on the grounds that the defeated candidate, Salvador Allende, was reputedly a communist. In December, the CIA conspired with the UK to whip up racial antagonisms in British Guiana (soon to be independent Guyana) and distort the election result in a way that kept the nation's founding father, Cheddi Jagan, out of office—successfully countering clandestine Cuban attempts to aid Jagan. Such actions, however necessary they might have seemed at the time, left the CIA open to the charge that it was instrumentalizing the United States' betrayal of its own principles of democracy and self-determination.

For after the Bay of Pigs, it was certain that people would suspect CIA interference in any politically unstable situation. In newly independent Ghana, the socialist president Kwame Nkrumah did not need proof of the CIA's courtship of his exiled opponents when he warned the US ambassador William P. Mahoney, "We've got to keep an eye on these people." When unrest in French Algeria welled up in 1963, French Foreign Minister Couve de Murville insinuated that there was CIA involvement, and President Kennedy felt uncomfortable in a Paris press conference, even if he opened in style—"I am the man who accompanied Jacqueline Kennedy to Paris"—for he realistically added, "The good will of this visit may be rapidly diminishing."[8]

Although covert actions continued, the Bay of Pigs did bring change to the CIA. President Kennedy decided he needed a scapegoat. The president told Dulles and Bissell they would have to go. This would be "after a decent interval." Perhaps that was no more than Dulles, so recently a guardian of the peace, deserved. Dulles stayed on through the morale-boosting completion of the CIA's new headquarters in Langley, Virginia. One historian has observed that Kennedy arranged for Dulles "to retire on a high note." The attribution of blame principle was, nevertheless, clear. Kennedy is famously reported to have told Bissell, "If this was the British government, I would resign, and you, being a senior civil servant, would remain. But it isn't. In our government, you and Allen have to go, and I have to remain."[9]

The departure of Allen Dulles on 29 November 1961 and of Bissell not long afterwards impressed the journalist Stewart Alsop as events that marked the end of an era. He felt that the days of his ilk were over.

With OSS in the war, Alsop had dropped into France to help the Resistance. He and his brother and fellow columnist Joseph were strong supporters of the CIA. Joe Alsop hosted many a dinner party in the 1950s at which senior CIA personnel were guests. Stewart Alsop wrote that the bureaucratic bloodbath that followed the Bay of Pigs meant the end of the times when graduates of private schools like Groton and Ivy League universities like Harvard supplied a steady stream of "Bold Easterners" who supplied the agency with imaginative leadership. Stewart Alsop looked back with nostalgia at "the time when the CIA was positively riddled with Old Grotonians" and viewed with regret the advent of a new generation he called the "Prudent Professionals."[10]

It is a complex issue. It has yet to be proved that there were more Grotonian types in the CIA than in other federal institutions. The myth did exist, and it survived. President Nixon complained about CIA people primarily drawn from the "Ivy League and the Georgetown set," and such comments have persisted down to the present day. At the same time, a shift from privilege to proficiency seemed to be occurring in the aftermath of Dulles's demise. The CIA's first use of Corona spy satellites went back to 1959, but they now carried improved high-definition cameras, and all of a sudden that kind of high technology seemed to epitomize the CIA's mission. Kennedy's choice of a successor to Princeton-educated Dulles added substance to the "prudent professional" counterimage. John McCone hailed from San Francisco, graduated from Berkeley with an engineering degree, and had served as the director of the Atomic Energy Commission under Eisenhower. According to the *US News and World Report*, he was "a man with a slide-rule mind."

Yet McCone was sufficiently "imprudent" to preside over Operation Mongoose. This was a renewed and intensified terrorist and economic sabotage campaign that Kennedy launched against Castro on 30 November 1961. Just when his brother Robert, as the US attorney general, was waging a high-profile war on the Mafia, the president authorized the CIA to reopen channels to the mobster killing machine with a view to disposing of Castro. The administration plumbed still deeper depths, for the CIA covert operations chief Richard Helms later stated that Robert Kennedy directly ordered him to operate through the Mafia. Meantime, the sexually insatiable president appears to have shown extra enthusiasm for such associations. He shared a "moll" with Sam Giancana, head of the "Chicago Outfit." The crooner Frank Sinatra was the intermediary who supplied the services of this serial mistress, Judith Campbell.

The FBI's J. Edgar Hoover told Kennedy he knew about this and kept the card up his sleeve, so McCone, along with other government officials, may have been in the dark about some of the more precise details. McCone was a Roman Catholic, a rather more serious one than the president. He reportedly fretted that he might be excommunicated should word slip out that he had sanctioned assassination as an instrument of state policy. In internal CIA exchanges that discreetly discussed the murder of Castro, his subordinates were enjoined to deploy judicious censorship. A memorandum drafted by William K. Harvey, who ran special operations in Cuba, requested the omission of the phrase "liquidation of leaders" from the record of a conversation involving Edward Lansdale. It is just possible that McCone remained in the dark about the more extreme designs implemented on behalf of the Kennedys by his own agency. If not, like CIA directors before and after him, he chose to comply and to keep mum.[11]

One of the most dangerous episodes in modern history, the Cuban Missile Crisis of 1962, soon tested the technological capacities of the CIA. Nikita Khrushchev ignited the crisis by installing nuclear missiles on the Caribbean island. There has been a tendency to denounce the Soviet leader for being rashly aggressive. The view from Moscow was different. Two maxims of Russian foreign policy had been important for centuries. They governed the Soviet Union as much as they had its tsarist precursor. The first was to ensure that the nation was never land-locked and had access to the oceans. The second was the desirability of borders that were secured by compliant, if not friendly buffer states. Invasions by the Mongols (Genghis Khan), Swedes (Charles XII), French (Napoleon), and most recently and tragically Germany (Hitler) had made national security a matter of extreme urgency in a way that few other countries could appreciate. When the United States decided to deploy intermediate-range Jupiter nuclear-armed missiles in Turkey within striking distance of Moscow, Khrushchev attempted humor, saying they were "aimed at my *dacha*." But he was angry.[12]

The Soviet leader is recorded as having said, "Since the Americans have already surrounded the Soviet Union with a ring of their military bases and various types of missile launchers, we must pay them back in their own coin... so they will know what it feels like to live in the sight of nuclear weapons." The Soviet military proceeded to install nuclear weaponry on Cuba by stealth, hoping to accomplish their aim before the US could make preemptive moves. By implementing this policy, Khrushchev breached America's own buffer zone protocol, the Monroe Doctrine, which the Soviet leader had already pronounced "dead."[13]

Cuba had conveniently become available as Moscow's Turkey, its "unsinkable aircraft carrier."[14] Kennedy's continuing campaign of assassination and disruption made Castro willing to accept military installations that would deter the United States from sponsoring further invasions of his own country.

The CIA's contribution to President Kennedy's handling of the problem of missiles that were within striking distance of Washington would be vital to the solution of the crisis. It was instrumental in averting nuclear catastrophe. The agency's performance saved and enhanced its intelligence stature and helped salve the Bay of Pigs humiliation.

In July 1962, the CIA noted an increase in the number of ships that put in at Cuban ports. It then spotted various earthworks and the emplacement of surface-to-air missiles of a defensive type. McCone had a hunch that the Soviets would go further and set up medium-range nuclear-armed ballistic missiles. He repeatedly warned about this. The White House would not listen, as he had no proof, and then the CIA director departed for Europe on an ill-timed honeymoon with his second wife. Perhaps because the Bay of Pigs attack had produced an upsurge of strong feelings in Cuba, the CIA's numerous informants on the island supplied a stream of manifestly unreliable information about all manner of threats. However, after the Gary Powers shoot-down, it was risky to send U-2 overflights, and high-definition satellite photography was not quite ready. Early warning of Soviet implementation would have given Kennedy more time to react, but the CIA mistakenly advised, in the absence of proof, that Khrushchev was innocent of evil intent.

Meantime, suspicions were growing, and the nation was expressing its support of its commander in chief: in mid-September, both houses of Congress assured Kennedy, by massive majorities, of their support for the use of any necessary force. A month later, the president risked a U-2 overflight, and the plane returned with low-resolution images of major works and machinery. Information arrived from Oleg Penkovsky, a colonel in the GRU, Soviet military intelligence, who had agreed to spy for British intelligence and would be arrested and executed by the

Soviet authorities the following year. It helped analysts interpret the photos, and the Kennedy administration thus learned that, as a matter of urgency, it would need to devise a strategy.

Kennedy refrained from ordering an attack on the Soviet weapons, as that might have precipitated an apocalyptic war. Instead, the president imposed a naval blockade or "quarantine" on Cuba. At a televised United Nations Security Council meeting in New York, US Ambassador Adlai Stevenson stunned his prevaricating Soviet counterpart by pinning up the conclusive pictures on a blackboard.

After a few hair-raising near-miss military incidents, Washington and Moscow resolved the crisis. In-house historians were able to portray Kennedy as the hero of the moment, and the sun had shone on the CIA. Khrushchev had shown maturity and courage. He backed down, explaining to his Kremlin comrades that he did not want a nuclear holocaust. A face-saving, if significant formula appears to have occurred to both leaders independently.[15] This was that the United States would withdraw its Jupiters from Turkey, and the Soviet Union its missiles from Cuba. Kennedy was able to retain the US naval base in Guantanamo Bay, Cuba, a potential bargaining chip, and the US Navy soon deployed a new generation of Polaris missiles that could be launched against Khrushchev's dacha without the inconvenience of being based in Turkey. Thus, the American public could be told that Kennedy won by surrendering only the soon to be obsolete Jupiter missiles.

Kennedy, through the CIA, continued his vendetta against Castro. In June 1963, the National Security Council authorized a new round of sabotage against Cuba, and the CIA's old Far East hand Desmond FitzGerald took on responsibility for assassinating Fidel Castro. One of his plans was to use an unhappy Cuban military official, Rolando Cubela, to administer poison via a syringe disguised as a ballpoint pen. CIA case officer Nestor Sanchez offered the weapon to Cubela at a meeting in Paris on 22 November.

That was the day that Kennedy died in Dallas, Texas. His assassin, Lee Harvey Oswald, had strong connections with both Cuba and Russia. The Warren Commission, set up to investigate the circumstances of Kennedy's death, included Allen Dulles, and, in the interest of national morale, stopped short of exploring some of the shadier possibilities. This reticence would feed into popular suspicions—suspicions that flared up when in the 1970s the American public first learned of the

CIA's own program of assassination. The idea that Castro ordered the assassination of Kennedy in self-defense became a preoccupation of later investigators. The House investigation of 1979 concluded that Castro was too rational to order such an event; yet the suspicion was sufficient to fuel the domestic turmoil of the 1970s, and numberless conspiracy theories since then. In the 1960s, however, Kennedy's death was generally accepted for the tragedy it was, and the CIA escaped, for the time being, the rage that would later descend upon it.[16]

Yet scandal was in the air, and however well connected the CIA may have been in the press, investigative journalism was alive and well in the United States. In March 1967 a West Coast radical Catholic magazine, *Ramparts*, published an exposé of one of the CIA's activities. Its author, Sol Stern, revealed that the agency had been secretly funding the US National Student Association's international activities. As the association was an American institution, the CIA had thus transgressed against the 1947 provision that it should operate only abroad. There followed a rush of revelations in the mainstream press about the CIA's suborning of teachers, academics, journalists, and labor unions. Press coverage was extensive: *The New York Times* alone published fifty-four stories about the *Ramparts* controversy.[17] Embarrassing discussions took place at home and abroad. Had such and such a person, hitherto respected for his/her independence, benefited from secret CIA largesse? If so, had it been knowingly or in an unwitting manner?

Congress awoke to the issue in a manner unseen since the 1940s. Kentucky's Carl D. Perkins, chairman of the House Education and Labor Committee, called for an investigation. His fellow Democrat Senator Mike Mansfield of Montana, who had been agitating for closer oversight of the CIA, complained that the CIA had never informed Congress of its subsidization of the voluntary sector. He denounced the policy as a "move toward big brotherism." Senator Barry Goldwater, a senior figure who had been the Republican presidential candidate in 1964, spoke for a conservative movement that was gaining traction. The Arizonan saw the scandal as confirming the liberal tendencies of an agency that had pursued an opening to the left. He condemned the CIA's effort "to finance socialism in America."[18]

The administration was caught unawares. Secretary of State Dean Rusk was out of town when the scandal broke, and it fell to his assistant, Nicholas Katzenbach, to handle the situation. Katzenbach's first instinct was to deny everything. The difficulty was that he did not know what he was denying: "I am not absolutely confident that we are [in] possession of all the facts." There was a hint, here, that the CIA was out of control, though the ignorance may have arisen from the State Department's traditional abstemiousness when it came to CIA matters. Compounding the issue was the fact that Kennedy's successor in the White House, Lyndon B. Johnson, was a domestic affairs specialist who was not always as curious as he might have been about foreign policy.[19] Perhaps President Johnson was truthful in letting it be known that he had been "totally unaware" of the National Student Association subsidies.[20]

Katzenbach set up an inquiry. Richard Helms, who had been appointed director of the CIA the previous year, was one of its members. The Katzenbach Commission considered but rejected the idea that America should establish an equivalent of the British Council that would cultivate support by openly sponsoring cultural events. Instead, it concluded: "It should be the policy of the US government that no federal agency shall provide any covert financial assistance or support, direct or indirect, to any of the nation's educational or private voluntary organizations."[21]

The plan, as in 1947, was that the CIA would promise to be good, but secretly be bad. Desmond FitzGerald circulated a dictum to all field offices saying that covert assistance should continue. At home, the agency pored over the *Ramparts* tax returns and leaned on advertisers to cancel their account with the magazine. It also launched a disinformation campaign against *Ramparts*, for example, by stating that it encouraged the practice of fellatio.

The Nation condemned the Katzenbach report as "piously expedient," but the furor began to subside. Helms later attributed this to what one might term the "flashback" tactic, the practice of pointing out that previous administrations of a different party, in this case the Republicans, had engaged in identical nefarious behavior. Democrats who deployed the flashback tactic included Georgia Senator Richard Russell, the powerful chairman of the senate Armed Services Committee, whose remit included CIA oversight. Another was former attorney general Senator Robert F. Kennedy, who backed Russell's statement that all presidents since the CIA's creation, including the Republican Eisenhower, had authorized clandestine subsidies.[22]

Perhaps more to the point, LBJ could afford to brush off the scandal because he was a Democratic president with substantial majorities in both houses of Congress. He still commanded the loyalty of his party—the discrediting of his administration by the Vietcong's Tet Offensive would not occur for another year. Maine's Margaret Chase Smith, the senior Republican on the Armed Services Committee, had sniped at the agency in the past, but was a close friend and ally of the president who described her role as that of "loyal opposition." As for the chief executive, Johnson was no great admirer of the CIA but, to apply a phrase he would make famous, he did not want it "outside the tent pissing in."[23]

Helms was the first director of the CIA to have come from an almost one hundred percent intelligence background. Before the war he had worked as a journalist in Germany (where he interviewed Hitler), but then it was OSS and CIA all the way. So in spite of his external brief, he could be insular. He assured LBJ's national security adviser Walt Rostow that the French foreign office, on reading the *Ramparts* stories, were impressed that the CIA had been "so clever" in its modus operandi. The CIA director refused to acknowledge the damage that knowledge of the exposed practices caused abroad.[24] He was fundamentally mistaken. Because the CIA's activities had taken place worldwide, they spread more widely the revulsion that the Bay of Pigs had set in train.

Foreign attacks on the practices of the CIA had been happening ever since the appearance of *The Cloak and Dollar War*, a 1953 work by the Australian communist Gordon Neil Stewart that drew on US press reports to discredit efforts to "roll back" communism in Eastern Europe. Like publications sponsored by the CIA that discredited communism, *The Cloak and Dollar War* could be dismissed by objective readers because of who paid for it. But it did herald a line of attack by communist writers who in the misdeeds of the CIA saw a point of American vulnerability, and who continued to attack what they saw as this soft US underbelly.[25]

The communists renewed their attacks on the CIA in the wake of *Ramparts*, and non-communist critics now joined them. At a time when Washington needed international support for its deeds in Vietnam, the committee rooms of the UN building in New York City began to reverberate to the sound of debate on the CIA. Chester Bowles had advocated the abolition of the CIA after the Bay of Pigs and was now US ambassador to India. He reckoned that *Ramparts* "hurt us throughout the world, particularly in India." In 1967 the US journalist Thomas B. Morgan undertook a tour of the world to assess the causes of anti-American feeling. His top three causes, in ascending order, were the Vietnam War, racial issues—and the CIA.[26] The Bay of Pigs fiasco did indeed have a long tail.

7

Vietnam: The Roles of the CIA

George W. Allen was a veteran of military intelligence when he joined the CIA in 1963. He gave his verdict on the CIA's subsequent performance in the Vietnam War in a book he published in 2001. According to Allen, there was "no consideration of objective truth" when he worked with a White House group dedicated to selling the Vietnam War to the American public. There was, he added, "surprisingly little concern about credibility."[1]

The historian Richard Immerman offers a different judgment. According to his account, the CIA pursued its objectives reasonably well in the course of the war. Its failure was more reputational than real, and that poor reputation was a consequence of the nation's failure to win the war.[2]

In the pages that follow, we shall see that there was a pessimistic tendency in CIA reports. If the assumption is that the war was unwinnable (and that is contested), the pessimistic tendency does come out looking good.[3] Yet at the time intelligence leaders did not speak up. If there were resignations over government policy, they were too low-key to make an impact. The CIA embraced the principle of "intelligence to please," and showed itself vulnerable to manipulation. Out of loyalty, the CIA helped the Johnson and Nixon administrations prosecute a war in which it did not believe, and killed people contrary to its own principles.

At the end of World War II, US policymakers saw Vietnam's nationalist insurgent leader Ho Chi Minh as a potential ally. Then, out of caution, they decided to back the reimposition of French colonial rule. Their caution turned out to have been rash. By 1954, Ho's forces had obliged France to give up half the country. The Geneva Accord of that year recognized the independence of North Vietnam, by now under communist control and styling itself the "Democratic Republic of Vietnam." Traces of waning French influence remained in the creation, south of a dividing demilitarized zone, of the separate Republic of Vietnam.

The United States had no vested business interests in Vietnam, and the country was of little strategic value, but Washington wanted to prop up the Republic of Vietnam (more commonly known as South Vietnam) in line with the domino theory. If adjacent South Vietnam became communist like the North, it could produce a reaction throughout Southeast Asia comparable to the ripple effect in a row of dominoes: if one falls over, so will the rest in rapid succession.

The CIA called on the services of Edward Lansdale. Needing cover, Colonel Lansdale became assistant US air attaché at the South Vietnamese capital, Saigon, 1954–6. He attempted to reapply the counterinsurgency formula he had developed in the Philippines. He boosted the image of Ngo Dinh Diem, prime minister and president of South Vietnam between 1954 and 1963, on one occasion arranging favorable coverage in *Life* magazine back in the United States. Washington had imposed Diem on his countryfolk, and Lansdale rigged an election to create the illusion that he had popular support. The CIA man arranged sabotage operations in the North, for example, by contaminating their fuel supplies. One of his psychological warfare tactics was to spread the rumor that the communist government based in North Vietnam's capital Hanoi was giving Ho the status of a deity and calling on citizens to worship him. Lansdale facilitated the migration of a million disenchanted Catholics from the North to the South. In South Vietnam they received favorable treatment from Diem, who was a Catholic and who had given preference to Catholics in his administration in spite of the overwhelmingly Buddhist composition of his nation's population.

Lansdale had some successes, but they were limited. The Viet Cong (Vietnamese communist) guerrilla movement began to gain ground in the South. Diem proved to be a flawed ally. He was an elitist conservative. He was furthermore a nepotist and frigid in his personal practices—he hated to be touched and refused even to wave to his people.

By the time Kennedy entered the White House in 1961, Lansdale had been back sitting at a desk in the US Department of Defense for a number of years. He would be an incessant source of ideas, for example, about how to kill Fidel Castro, but for the time being took a

back seat in regard to Vietnam. The Kennedy administration cast about
for new inspiration. Although the French had long experience of
fighting communism in the region, America ignored them and turned
to a British military expert.[4] Colonel Robert G. K. Thompson had
orchestrated the Malayan government's response to the "Emergency,"
a communist guerrilla insurgency. British–Malayan tactics had included
placing loyal populations in fortified villages, winning "hearts and
minds" by supplying food and security. The policy was also to shoot
everything that moved outside the villages—the peninsula's monkey
population took a hit, and by 1960 the communists had been wiped out.

Thomson arrived in Washington. After one meeting with the CIA's
Far Eastern specialist, Desmond FitzGerald, the British officer recalled,
"We devised names for each other so that we could talk more freely
on the phone. Desmond was Sanders, I was Carruthers."[5] Diem soon
became a Thompson enthusiast. The CIA trained the military cadres
entrusted with the defense of what were intended to be Malayan-style
fortified villages. By March 1962, it was able to report the creation
of 2,000 "strategic hamlets" in South Vietnam. By the same month in
the following year the figure was 5,000. At this point, Thompson
boasted of "impressive" progress and claimed that victory was round
the corner.[6]

The Thompson plan was doomed to failure. Vietnam was not a peninsula. The Viet Cong created and used the circumventing Ho Chi
Minh Trail. Based on ancient trade paths through the jungle and invisible from the air, the Trail snaked its way out of North Vietnam into
neighboring Laos and then Cambodia, bypassing the Demilitarized
Zone that separated North and South Vietnam. Terminating in South
Vietnam's Central Highlands, it was a means of resupplying and rein-
forcing the Viet Cong. By the later stages of the war, around 20,000
North Vietnamese regular soldiers traveled the highway monthly.

Realizing that Thompson's claims were spurious and blaming Diem
for the demoralized state of the South, America's ambassador in Saigon,
Henry Cabot Lodge, Jr., resorted to another type of action. Disregarding
the views of CIA director McCone, Lodge gave the nod to local plotters, who, on 2 November 1963, assassinated their president.

Political instability ensued. But what really killed off the nation-
building counterinsurgency approach was America's resort to overt
military action. President Lyndon B. Johnson decided to use heavy
bombing and ground troops with the intention of defeating the

combined efforts of the Viet Cong and the North Vietnamese army. His decision meant it was impossible to avoid "collateral" damage, the loss of Vietnamese civilian life and property. The resultant bombs and bullets were incompatible with the aim of winning indigenous "hearts and minds."

It was not that Johnson was a warmonger by nature. He desperately wanted to be able to concentrate minds and money on domestic reforms such as civil rights, the elimination of poverty, and the provision of universal Medicare. Ignorant of foreign affairs and impatient to end the nagging difficulties in Vietnam so that he could tend to his true love, he turned to physical force, intending it to be a quick fix. Tragically for him, his decision diverted the resources he needed for domestic reform and ended in defeat.

At the instigation of Secretary of Defense Robert McNamara, Pentagon officials collected, in the years 1967–9, documents that would amount to a thorough history of America's involvement in Vietnam. The resulting 7,000-page compilation came to be known as the "Pentagon Papers." In the words of historian John Ranelagh, the papers showed that the CIA's predictions about the outcomes of US policy on Vietnam had been "consistently pessimistic."[7] CIA officials carefully nursed that reputation, and by and large it was deserved. However, the agency's leaders did not, when it mattered, express their doubts in a forceful manner. In effect, they supported LBJ's clumsy policies when it would have been wiser to oppose.

The CIA contributed accidentally to America's virtual declaration of war, as did its partner in secret work, the National Security Agency, established in 1952 to coordinate US signals intelligence. Between them, the two agencies gave plausibility to the rationale behind the Gulf of Tonkin Joint Resolution of 7 August 1964, whereby the two houses of Congress authorized the president, by overwhelming majorities, to take "all necessary measures to repel any armed attacks against the forces of the United States."

The resolution was a reaction to a reported attack on the US destroyer *Maddox*. The *Maddox* had been conducting military surveillance operations in the Gulf of Tonkin, off the coast of North Vietnam. It was believable that the North Vietnamese had become sufficiently jumpy to mount an attack, as the CIA was conducting sabotage operations against bridges, storage dumps, and other targets in North Vietnam. In the event, the crew of the *Maddox* panicked and fired into the maritime mists at an enemy that existed only in their imagination. According to the National Security Agency's official historian, Robert J. Hanyok, the signals agency later concluded, on the basis of a carefully considered *ex post* analysis of intercepts, that there had been no attack on the *Maddox* (Hanyok completed his historical report at the time of the Iraq War controversy in 2002, and its release was delayed for three years, as comparisons between the two events would have been sensitive).

That proof of innocence was missing from the cryptanalysts' contemporary report on the Tonkin event, apparently lost in translation. The truth did not get through to the White House. Johnson believed he had the justification he was waiting for, and Congress obliged by authorizing him to conduct an undeclared war. By the end of 1965 there were 185,000 US military personnel in Vietnam. The resultant fighting would cost the United States 58,226 lives, and Vietnam one million dead.[8]

On 2 March 1965, the US Air Force started Rolling Thunder, a massive bombing campaign aimed at the North Vietnamese infrastructure. It would last for years and would bring about a great deal of death and destruction. Yet Rolling Thunder had a limited impact on the enemy's capacity to fight. For the selected targets were industrial, and North Vietnam was still a predominantly rural economy. Six days after the launch of Rolling Thunder, on 8 March, the US Marines landed in South Vietnam's Da Nang, with the purpose of defending the air force base in that coastal city. It was the first commitment of US troops to ground fighting and meant the war was well and truly under way.

The CIA was involved in these decisions to bomb and to fight, but in an indecisive way. CIA director John McCone had been a Vietnam pessimist in the Kennedy administration, but changed his mind, or at least his posture, to support the president's efforts to suppress the Viet Cong. In 1964–5, his analysts questioned the benefits of the proposed bombing campaign. Some evidence, for example, from the London Blitz of 1940–1, suggests that bombing actually increases the morale of the targeted people. Bombardment was also against the CIA's Lansdalian philosophy.

McCone failed to articulate such reservations to the president. However, Johnson rightly sensed that the CIA director was not enthusiastic about his decisions and froze him out of his influential Tuesday luncheon policy meetings. McCone resigned, to be replaced by his fellow Texan, Admiral William Raborn, a man said to be programmed to say "Aye, aye, Sir."

In his last attendance at the National Security Council, McCone did predict that by following the Pentagon's escalation plan the US "would drift into a combat situation where victory would be dubious and from which we could not extricate ourselves."[9] He claimed in later years that he had been "unhappy" about the military escalation, "and then when Johnson...agreed with the recommendation of McNamara that he put our troops on the offensive, that is when I parted company with them."[10]

McCone's deputy director for intelligence, Ray Cline, made a similar claim, saying he "tried to warn that an Asian guerilla war was not to be easily won by conventional military forces and weapons." He did not want to be responsible for "a tragedy for the United States and the peoples of Southeast Asia."[11]

In March 1966, Cline emulated McCone in stepping down. In his memoir he gave the strong impression that this was because he was unhappy with Rolling Thunder. Yet at the time Cline had voiced his support of air strikes. Furthermore, his resignation fell short of being a protest, as he accepted the post of adviser to the US ambassador in Bonn. It seems that he quit, in part, because he was running short of friends in high places. He was reportedly running a sniping war against his new director, Raborn, that made him enemies. National Security Adviser McGeorge Bundy was no friend and remarked on Cline's "obsessive personality." But Cline stopped short of offending LBJ. He reassured the president of his strong support, "especially in facing up to the tasks and dangers before us in Southeast Asia."[12]

In 1969 Cline would return to the United States when President Nixon appointed him director of the Bureau of Intelligence and Research (INR) within the State Department. Ever the sniper, he took issue with National Security Adviser Henry Kissinger. In March 1971 he complained to Kissinger that his advice had been overlooked with regard to the recent Lam Son 719 campaign. This was a US-backed South Vietnamese invasion of Laos with the object of blocking the Ho Chi Minh Trail, a military incursion that was frustrated, in the event, by superior North Vietnamese intelligence. Cline's second complaint had a familiar ring—the CIA and the Defense Intelligence Agency were competing rather than pulling together. Finally, Cline made a plea for net estimating, so that strategy could be informed by data on both enemy and US forces. Prohibited from domestic work, the CIA could not produce net estimates, but Cline suggested his bureau could. Kissinger was deaf to any plea that diminished his own agency in decision-making, but Cline's complaint was both a thoughtful critique of the US intelligence system and, in being made but not followed up, a further illustration of how senior officials could be critical but at the same time ineffective.[13]

To return to Rolling Thunder, the two CIA officials who were most authoritatively equipped to say no to the president, and perhaps to resign in protest against his policy, did not do so. Over the years, the usual defense of such inaction has been that the CIA is apolitical. But in practice the inaction led to the politicization of the CIA's product, with political, military, and humanitarian consequences that were, all agreed, tragic.

With the war showing no signs of abating, President Johnson redeployed Lansdale in 1965. Lansdale attempted to win over the ordinary people of South Vietnam, but with little success in the face of mounting civilian casualties. In March 1967, the military command launched a program called "Civil Operations and Revolutionary Development Support" (CORDS), an attempt to "pacify" the villages on principles far removed from those of the counterinsurgency maestro, who in 1968 quit, disillusioned.

Meantime, trouble was brewing at home. Public opinion had at first favored the war, but by 1967 it was wavering. Students objected to the draft. Black Americans, while keen to serve in the early stages of the war, were in revolt over both civil rights and, increasingly, Vietnam. Johnson, who rightly thought of himself as a champion of social reform, could not understand this. He took refuge in a conspiracy theory, the idea that domestic protest was not spontaneous but orchestrated from Havana, Beijing, and Moscow.

Johnson pressured the CIA to set up a special operations group with the task of finding out the "extent to which the Soviets, Chicoms [Chinese communists] and Cubans are exploiting our domestic problems in terms of espionage and subversion." Deputy Director for Operations Thomas Karamessines warned counterintelligence chief James Angleton that the program, being domestic, was illegal. Regardless, the group launched Operation Chaos and compiled files on 7,200 US citizens. Student leaders like Tom Hayden came under surveillance. On 15 November 1967, CIA director Richard Helms, who had succeeded the floundering Raborn in June 1966, reported to the president. The CIA had found "no evidence of any contact between the most prominent peace movement leaders and foreign embassies." He enclosed a report that described Hayden and his colleagues as "tireless, peripatetic, full time crusaders." The CIA had on this occasion spoken truth to power and had begun the foreign-policy education of Lyndon Baines Johnson.[14]

The Tet Offensive discussed below would brutally complete that education, and it stemmed from intelligence failures. One of these was linguistic deficiency. As in the case of the Soviet Union, there was an urgent need to determine the mindset and intentions of the communists in North Vietnam. Here there was a need for Vietnamese speakers, but they were in short supply. Lansdale's friend William Lederer had flashed a warning light about this problem in an appendix to his co-authored novel *The Ugly American*. Making a comparison with Soviet foreign representation, Lederer observed that nine out of ten Russians spoke the language of the nation to which they had been assigned. Only 50 percent of US foreign service personnel spoke any foreign language whatsoever.

Head of the CIA's Far East Division by 1963, William Colby sent agents to spy on North Vietnam. His "Project Tiger" had the CIA using South Vietnamese with whom its officers could barely communicate, and who were open to offers from both sides. Scores of the agents engaged never returned from North Vietnam. In 1995, Captain Do Van Tien went on Hanoi television and explained one of the main reasons why. Do Van Tien had been the deputy chief of the indigenous Tiger unit run by Colby, but he had been a spy for Hanoi all along. Colby realized that Project Tiger was a disaster. He terminated it, and after that North Vietnamese strengths, weaknesses, and intentions had to be estimated—or guessed at—by deductive methods.[15]

General William C. Westmoreland's Military Assistance Command, Vietnam (MACV) was able to estimate the number of regular troops in the North Vietnamese armed forces, and he made deductions from that. He maintained that by increasing the number of US troops he could reduce the number of soldiers at the command of the North through combative attrition. Hanoi would have to limit the scope of its operations and would finally lose the war.

A young Harvard-educated analyst working alone at the CIA took a different view. Sam Adams counted guerrilla-militia forces as well as the regular troops infiltrated from the North. At the end of 1966, the US armed forces chief gave enemy strength as 270,000, but Adams made it 600,000. Adams's reports were withheld from the public and ignored. In his contention, that was because Westmoreland favored lower estimates of enemy strength to justify his claims of progress and to bolster his constantly swelling requests for US escalation—American armed strength in Vietnam would reach a peak of over half a million men.[16]

The Tet Offensive of 30–31 January 1968 was a stunning blow for President Johnson and his fellow Americans that at first sight justified Sam Adams's estimates. A joint force made up of southern irregulars as well as North Vietnamese troops attacked South Vietnamese and US targets in what was supposed to be the Tet holiday's truce season. The force's commanders committed 84,000 men to battle. Twenty-seven out of forty-four provisional capitals came under fire, and so did the American Embassy in Saigon. Da Nang was one of a number of urban strongholds to fall into enemy hands.

Then came the counterattack. The South Vietnamese armed forces lost 2,300 men fighting to repel the combined forces of North Vietnam and the Viet Cong, and 1,100 Americans died. However, the communists lost almost half their men, with 40,000 falling in battle. They failed in their objectives of overthrowing the Saigon government and prompting a general rising of the South Vietnamese people. It was a crushing defeat in military and political terms.

Yet it was a psychological victory for Hanoi and its allies. The Tet Offensive shocked America and especially President Johnson. According to Westmoreland's logic, the communists should never have been strong enough to launch such an operation. Understanding that his credibility was destroyed, Johnson decided not to run for re-election in the fall of 1968, and all the talk from now on was about how the United States could withdraw with honor.

When Richard Nixon entered the White House in 1969, he had that objective in mind, but withdrawal would take another six years as the Republican president strove in vain to exit without outright defeat. The CIA's song of pessimism continued. At the same time, the agency operated ever more widely, for example, in the neighboring nations of Laos and Cambodia, and in Vietnam itself it embarked on Operation Phoenix.

Operation Phoenix took place in the context of Nixon's policy of "Vietnamization," a bastardization of the Lansdalian policy of trusting, training, and arming indigenous peoples. It could not be separated from the deployment of much greater use of force than Lansdale had used in the Philippines. Notably, the bombing campaign Rolling Thunder continued. But Vietnamization had a special political attraction for President Nixon. Student protests against the draft had reached a crescendo and now he would be able to let the young men of Vietnam do the fighting, while American boys stayed at home.

Phoenix ran from 1968 to 1972 and was a CIA-led effort to help the Army of the Republic of Vietnam rid South Vietnam of communism by the direct method of killing communists. The idea was to identify civilian Viet Cong leaders and activities in South Vietnam's villages and then to neutralize them. Suspects were interrogated, imprisoned, and killed. Men and women were tortured physically and sexually prior to death. The South Vietnamese were entrusted with implementation, and that led to some false accusations against individuals for extraneous reasons, such as family feuds.

William Colby administered this program of torture and murder. Restraint was not his forte. When with the OSS, he had commanded a unit that captured a German patrol in Norway and then wiped it out.[17] He had more recently had a hand in the execution of over a million suspected communists in Indonesia. Against that background, the assassination of 20,000 actual and alleged Viet Cong must have seemed relatively modest. However, the program provoked a strong reaction. It helped to solidify hatred of the Saigon regime, which finally fell to the communists in 1975.[18] The CIA had predicted defeat in Vietnam and then contributed to it.

At the end of the 1960s, the CIA still enjoyed a high standing with many Americans. Students who objected when its recruiters appeared on campus were in a minority. Together with the FBI, the CIA remained one of the few government agencies that enjoyed iconic status. The general public were unaware of the faults in America's intelligence system that had resulted in the quashing of McCone's reservations and Adams's calculations. The *Ramparts* revelations of 1967 had lifted one corner of the veil, but people had no inkling of the agency's illegal surveillance of domestic protest groups and were unaware of the agency's assassination programs ranging from Mongoose to Phoenix.

In the White House, however, there was scant respect for the agency. LBJ had not been a fan, and Nixon was downright suspicious. Nixon had humble origins. According to Henry Kissinger, he "considered the CIA a refuge of Ivy League intellectuals opposed to him." When the CIA failed to warn the president of the destabilizing fall of the Norodom Sihanouk government in Cambodia in 1970, he complained about the "clowns... out there at Langley." He was under no illusion about the nature of some of the agency's work: "This organization has a mission that runs counter to some of the very deeply held traditions in this country and feelings, high, idealistic feelings, about what a free society ought to be." The low standing of the CIA in the White House would have far-reaching consequences, especially when the public turned against the agency too.[19]

8

From Reformation to Counter-Reformation in the 1970s

A multipronged political attack endangered the CIA's standing in 1975. The attack resulted in reforms. Though controversial, these reforms contributed to a restoration of popular faith in the CIA. The attack on the CIA occurred for a number of reasons. We can begin with the distress experienced by a single individual, Catherine Colby.

Catherine was the daughter of William Colby, the architect of Operation Phoenix who would serve as director of the CIA, 1973–6. Like other officials at the time of the Vietnam War—for example, defense secretaries Robert McNamara and Melvin Laird as well as Nicholas Katzenbach—Colby found no rest at home when his offspring awoke to the nature of his work.[1]

Catherine suffered from depression. To put himself in a position to try to help her, in November 1971 her father left his field position in Vietnam to take up a CIA desk job in Langley. Part of the youthful rebellion against the Vietnam War and upset about Phoenix, Catherine could not rid her fevered imagination of images of suffering Vietnamese villagers. Her father tried to calm her but failed. In 1973, Catherine would take her own life.[2]

To some degree because of this tragedy, Colby embarked on a quest for redemption. He would write autobiographical stories for his grandchildren in which he sanitized unsavory aspects of his past, a common practice for grandparents, yet also an undertaking that pointed to a troubled conscience.[3] All this is significant because it helps to explain why Colby cooperated, to a considerable degree, with congressional investigators, allowing them to uncover dirt in a way that intensified the debate about the agency and its role. However, that cooperation did have an element of calculation. Colby was a tactician who knew how to retreat in order to advance.

A second circumstance contributing to the "intelligence flap" was an outbreak of US–USSR harmony. Peace initiatives were attractive to the Nixon administration in light of America's weakened military strength. In the course of the dollar-consuming Vietnam War, the United States had slowed down its spending on nuclear defense systems. Meantime the CIA, all too obligingly in the eyes of some American hawks, had scaled back its estimate of the Soviet military threat. Still, it was evident the balance of military power did seem to have changed in favor of the USSR. By the time President Richard Nixon's foreign-policy maestro Henry Kissinger assumed his responsibilities as national security adviser in January 1969, it was in the interest of the United States to seek an arms limitation deal with the USSR.

Kissinger's "linkage" diplomacy sent a double message to the Soviet Union. Nixon broke with Cold War custom and visited communist China in February 1972, while Kissinger sought peace in Vietnam and secured a withdrawal agreement at the Paris Peace Accord of January 1973. The CIA helped organize the final evacuation of Saigon with dramatic scenes of personnel scrambling at the last minute up a ladder to access a rooftop helicopter, a chaotic retreat that was at first sight humiliating but was to America's advantage in dealing with Moscow. An America reconciled with China and free of the draining Vietnam struggle gave the appearance of being able to outflank and outstrip the USSR. Moscow read Kissinger's implicit message: it was time to make peace with the United States.

The United States and the Soviet Union signed, on 26 May 1972, an existential pact. This was the Strategic Arms Limitation Talks Agreement. As it had a successor, the agreement came to be known as SALT I. One of the greatest diplomatic achievements of the modern era, SALT I stabilized a nuclear ratio between the United States and the Soviet Union, and limited the deployment of missiles with multiple nuclear payloads.

The CIA played a humiliating role in SALT I. Kissinger and Nixon, while eager to deploy the agency's covert operational capabilities, had a cavalier disregard for the integrity of its estimating function. In a calculating fashion, they defamed the agency's estimative capabilities and invented opportunistic intelligence in their place. First, they argued that there was a Soviet weaponry buildup that demanded the construction of a US anti-ballistic missile system. It was a bluff to bring Moscow to the negotiating table. Next, Kissinger and Nixon massaged the truth in order to persuade the Senate to ratify the resultant agreement. CIA data showed that the Soviet Union had a track record of cheating on agreements. Kissinger suppressed that worrisome information, and the Senate ratified.[4]

The national security adviser had wrought a masterly deception that boosted US national security and self-confidence. It had two challenging consequences for the CIA. First, Kissinger had reduced the agency's estimative function to a political plaything. Second, the outbreak of peace contributed to a change in political mood. Feeling more secure, voters, media personnel, and legislators would set aside their normally instinctive loyalty and, once they learned of its infractions, would feel free to criticize the CIA.

Watergate was a third leading cause of the CIA's political ordeal. The infamous affair led to the resignation of President Nixon, to a widespread tendency to distrust and question federal government, and to a besmirching of the reputation of the CIA. The affair centered on the exploits of the "Plumbers." This was the informal title of a "special investigations unit" that White House staff assembled in order to plug embarrassing leaks. The group included two Cuban Americans and Frank Sturgis, the supplier of Marita Lorenz's poison pills. It operated under the leadership of E. Howard Hunt, a former CIA operative who had written a series of indifferent novels in an attempt to boost the agency's mystique.[5] Hunt and his Cuban Americans had participated in a further failure, the Bay of Pigs invasion.

The Plumbers came into being because, in June 1971, the US military analyst Daniel Ellsberg had leaked portions of the Pentagon Papers, via journalist Neil Sheehan, to the *New York Times* and *Washington Post*.[6] The Papers, in edited format, having been reduced from the 7,000-page original commissioned by Robert McNamara, attributed to the CIA a conspicuous degree of skepticism regarding the Vietnam War. They thus added fuel to flames of dissent that were already licking at the door of the White House. So the administration authorized the Plumbers to break into the premises of Ellsberg's psychiatrist in a (futile) search for dirt.

A more infamous enterprise lay around the corner. In June 1972, the Plumbers broke into the offices of the Democratic Party situated in Washington, D.C.'s Watergate complex. The mission reflected the insecurity of an otherwise competent chief executive, for according to a contemporary gibe the burglars were unlikely to find anything more compromising than evidence of the Democratic Party's dire financial position. Had Nixon pleaded mea culpa, the Watergate break-in would have faded into insignificance. Instead, he ordered the CIA to obstruct the inevitable FBI investigation into the crime. Helms refused. Nixon already bore Helms a grudge. Under his orders, Helms had renewed the CIA's investigation into the US protest movement and once again ruled out foreign conspiracy. Now, Nixon fired Helms.

James Schlesinger was his replacement. Becoming privy to the history of the Plumbers and not wanting to be caught off guard by further disclosures in the future, Schlesinger launched a preemptive inquiry into all crimes and indiscretions that might be laid at the door of his agency. The resultant report contained what came to be known as the "Family Jewels"—a list of hidden CIA scandals that would burn its way into the open and play a role in the forthcoming flap.[7]

By the time this happened, the public had feasted on exposés of Watergate in the press. Bob Woodward of the *Washington Post* became a household name because of his Watergate stories. The Senate held public hearings. Fearing impeachment, Nixon quit the White House on 9 August 1974. Public respect for Washington government in general reached a low ebb. The CIA was a focus of suspicion because of Hunt's role in the Watergate break-in, but at this stage only in a minor and indirect way.

The timing of a fourth cause of the CIA scandal was unlucky for President Gerald Ford. He assumed office on Nixon's resignation, and on 8 September 1974 issued a pardon to indemnify any crimes his predecessor might have committed when president. The idea was to draw a line under Watergate and to move on. But, on that very day, Congressman Michael J. Harrington stirred up a new storm. It was about Chile. In 1973, a military coup had overthrown the democratically elected President Salvador Allende, and the head of state had died in questionable circumstances in the siege of the presidential palace. General Augusto Pinochet had seized power and showed early signs of developing into a vicious dictator. Harrington supplied the *New York Times* and *Washington Post* with text referring to William Colby's confidential testimony to a congressional subcommittee. Colby had mentioned the CIA's involvement in Chilean politics. Harrington now demanded an investigation into the CIA's involvement in the death of Chilean democracy.

Unknown to the American public, the CIA had (along with its Soviet and Cuban counterparts) immersed itself in Chilean politics since the early 1960s. It applied well-tried, if somewhat unproven methods, in this case secret finance for conservative parties and black propaganda to discredit the left, in particular the socialist Allende. Under both LBJ and Nixon, presidential direction of the agency in these matters indicated a long-running blind spot in American foreign policy. Whether from fear or from material motives, the opening to the left was now dead, or at least not applicable outside Europe. Non-European peoples were not allowed to choose what so many of them wanted, a little bit of socialism and a little bit of capitalism, together with democracy.

When Allende finally won election in September 1970, Nixon directed Helms to try for a coup to prevent his inauguration. When that failed, Helms passed on Nixon's wrecking order: "$10,000,000 available, more if necessary—full time job—best men we have—game plan—make the economy scream."[8] And on 22 October 1970 there occurred the assassination of the Chilean army's commander in chief, René Schneider. It was another attempt to destabilize the left-wing government of Chile. Helms congratulated local CIA officers on their "excellent job" of inspiring the surrogate killers.[9]

Once people learned of these details, it did not take a leap of imagination to think of the CIA as being up to its elbows in blood and criminality. When President Ford responded to the revelation, saying that Allende, not the CIA, had been the threat to Chilean democracy, Minnesota's Senator Walter Mondale retorted that was "hogwash." His fellow Democrat, Congressman Harrington, asserted that the CIA, not the State Department, was running US foreign policy and wrote to Ford saying that the agency's covert operations in Chile would turn world opinion against the United States.

Domestic opinion was another matter. While a Harris poll found that a mere 18 percent of those asked supported the Chilean intervention, only 29 percent opposed it, a result that pointed to a typical widespread apathy toward faraway events.[10]

The regime change in Chile did, however, feature in the rhetoric of an emergent and potent type of critic of US intelligence practices, the apostate. Victor Marchetti was an influential example of the genre. From a working-class background in the coal communities of Pennsylvania, he had joined the CIA and at the peak of his career enjoyed the privilege of a morning coffee routine with Director Helms. Then he became disillusioned, resigned, and in June 1974 joined with John Marks, a veteran of State Department intelligence, in publishing the book *The CIA and the Cult of Intelligence*. Marchetti and Marks argued that, through the CIA, America had polluted its mission to foster self-determination. The CIA had become obsessed with covert operations. Its machinations had facilitated the coup in Chile.[11]

Complying with the terms of his employment contract with the CIA, Marchetti had submitted drafts of his book for official clearance. Losing his patience over the agency's objections to over 300 passages in his draft, he and Knopf, the publisher, went ahead and printed the book with 168 spaces, and a further 141 passages printed in bold, highlighting statements that the censors had originally objected to and then reluctantly restored. For example, thirty-six lines of the section dealing with Chile were redacted, and one of the restored sentences printed in bold stated, "The 1968 [intelligence] estimate had in effect urged against the kind of intervention" that occurred in Chile.

The publishers exploited this censorship. The front cover of the paperback version trumpeted "the book that the CIA tried to suppress." Marchetti became a bestselling author. He and other CIA renegades like Patrick McGarvey, Sam Adams, and Philip Agee contributed to a wave of discontent that resonated with a virulent revisionism then sweeping the nation's campuses. Previously attacked for its alleged left-wing tendencies, the CIA now became a target of choice for the "New Left." Historians like William Appleman Williams and Gabriel Kolko had challenged self-validating narratives of the Cold War, emphasizing not the sins of communism but the hypocrisy behind a business-driven and expansionist US foreign policy that disguised itself as a liberating force. Their works inspired a new and critical generation on the nation's campuses.[12]

On 17 December 1974 CIA director Colby had a meeting with James Jesus Angleton. The agency's counterintelligence chief, Angleton enjoyed legendary status. He was a patriot, Cold War-style, who collected conspiracy theories in the way that others collected postage stamps, and believed it was better to be suspicious than sorry. Sane colleagues regarded him as paranoid for believing that the widening

Sino-Soviet split was a deception ploy aiming to seduce the West into lowering its guard. Judged by results, though, Angleton was a success. He had kept the CIA free of Soviet penetration in a way that Britain's MI6 could only envy. The time had come, Colby told this paragon of national security, to resign. The reason? Journalist Seymour Hersh was about to publish an exposé of continuing CIA domestic surveillance during the Nixon administration. Because he had overseen that program, Angleton would be a prime target.[13]

Hersh's front-page *New York Times* story of 22 December 1974 was authoritative. He had already won respect and a Pulitzer Prize for his reporting of the My Lai war crime of March 1968, when two companies of US infantry embarked on an orgy of raping and killing in two Vietnamese villages that resulted in the deaths of 400 to 500 men, women, and children. Hersh had verified some details of the CIA's surveillance activities with Colby. For the CIA director was pursuing a policy of controlled transparency, a decision that dismayed Angleton, who darkly hinted Colby might be working for "the other side."[14]

Although there were precedents, not least the abuses revealed in *Ramparts* in 1967, Hersh concentrated on the infractions of the Nixon administration, an already wounded target in the wake of the Watergate scandal. He told of the CIA's illegal surveillance of antiwar protesters and other dissidents, and detailed some of the methods used, including mail opening, telephone tapping, and burglary. American opinion had long been hostile to the use of such practices at home and was now primed to turn on the CIA.

Christmas 1974 had not yet arrived, but the "Year of Intelligence" had already begun. In the calendar year 1975, 137 US newspapers with a combined readership of 28 million would make the CIA the focus of 227 editorials. It was a dramatic shift of focus. In the year 1970 there had been no such editorials.

The Hersh revelation sprang from the same hymn book as the Hughes–Ryan Amendment to the Foreign Assistance Act of 30 December 1974. That new law required the president to report covert operations in advance to relevant congressional committees. It was a clumsy piece of legislation, as it meant that up to 170 legislators would have to be briefed, offering the prospect of a secret agency that leaked like a sieve. But it was the start of a reform process aimed at curtailing such travesties as the US snuffing-out of democracy in Iran, Guatemala, Chile, Guyana, and Indonesia.

In the early months of 1975, the White House responded to the growing crisis of confidence in the CIA. Four days into the new year, President Ford established a "blue-ribbon" commission on CIA activities within the United States under the chairmanship of Vice President Nelson Rockefeller. Eight members of the American establishment sat on the Rockefeller Commission, and Ford hoped for a whitewash. The prospects of such an eventuality dimmed when the US Senate on 27 January set up a select committee to study governmental operations with respect to intelligence activities, generally known as the Church Inquiry after its Democrat chairman, Senator Frank Church of Idaho. Then, on 19 February, the House of Representatives set up its own select committee on intelligence. Otis G. Pike, another Democrat, if a more conservative one, came to chair that committee and lent it his name. Whereas the Church Committee looked at intelligence abuses across the board, the Pike Committee asked questions about the effectiveness of US intelligence. The denizens of Langley were in for a comprehensive examination.

Now there occurred a revelation that shook the American psyche to its core. Pursuing his policy of disclosure in the interest of expiation and closure, Colby had shown the Family Jewels to President Ford. Following this, at a White House luncheon for senior editors, Ford was explaining how the Rockefeller Inquiry would be responsibly discreet about what otherwise would be dreadful disclosures, when he was thrown off his guard. "Like what?" asked one editor. "Like assassinations," the president let slip.[15]

On 28 February, one of the journalists who learned of the event, Daniel Schorr, leaked the news in a CBS broadcast. The nation was horrified, for in recent years it had been traumatized by the assassination of some of its leading citizens: former attorney general Senator Robert F. Kennedy, Martin Luther King Jr., Medgar Evers, and Malcolm X. Above all, the inadequacies of the Warren Inquiry into John F. Kennedy's death, combined with conspiracy theories pushed out by the KGB, had already created suspicions as to why America's president had been targeted on that fateful day in Dallas, Texas. Now, the possible motives were there for all to see. As Rockefeller and Congress inevitably turned to investigating the issue of assassination, a few of the details of the plots about Castro and other foreign leaders were exposed. Could Kennedy have been murdered out of revenge or preemption? Killing people abroad all of a sudden had a vivid domestic flip side, asking America to think again.

As the year 1975 unfolded, the three inquiries went about their work. The Church investigation was the largest Senate inquiry of the twentieth century and commanded a great deal of attention in the media. On 15 May, the committee heard from Colby in a closed hearing. But there were also extensive open hearings that held the public's attention through being shocking and at times gruesomely entertaining. On 16–18 September, for example, witnesses testified about the CIA's stockpile of shellfish toxins.

John F. Kennedy slipped from his perch as a liberal icon. Americans learned that he suffered from venereal disease. There came further disillusionment with the revelation that the Mafia moll Judith Campbell had numbered among his mistresses, news that fueled speculation about White House–Mob conspiracies to kill Fidel Castro. Four days after the Campbell disclosure, the Church Committee released its report on the CIA's assassination programs. Quite apart from damaging America's reputation abroad, confirmation of the existence of those programs had two effects at home. It shocked a public already sensitized to sudden death by the tragic loss of some of its own leaders. And, according to *Newsweek*, the issue caused the Church Committee to divert resources and attention from the domestic infractions that had "triggered the CIA flap in the first place."[16]

Be that as it may, the Church Inquiry into foreign excesses was also limited. Finite time and resources played their part in explaining that. After all, the United States was a global power. An investigation into every CIA undercover operation would have amounted to a research project into the recent history of practically every nation on the planet. Another reason for limitation was that the committee struck a bargain with Colby. The CIA director would protect the majority of his secrets but pass across information on one case that the committee could examine in depth for the indications that it might give about general practice. That case was Chile, and a detailed report of the CIA's efforts to stop socialism and overthrow democracy in that nation was published on 18 December 1975.

Party partisanship and image manipulation featured in the "season of inquiry." President Ford offered transparency, but of a limited type. His reticence stemmed from the fact that he was a Republican faced with confrontational Democratic majorities in both houses of Congress, and that he felt the need to defend the record of fellow Republican President Nixon. He also wanted to protect executive powers and—in the interest of national security—the principle of secrecy.

Ford's Rockefeller Commission was meant to appease the CIA's increasingly numerous critics. It failed in that objective. When it reported on 11 June, it recommended that there should be a single joint congressional committee to oversee the CIA, which would thus reduce the possibility of leaks. It acknowledged there had been domestic abuses but said little about assassinations and nothing about the possibility of a Castro revenge motive. Documents finally released in 2016 revealed that Vice President Rockefeller had bowed to administration pressure. Deputy White House chief of staff Richard Cheney recommended against publication of the Rockefeller Report and then helped ensure that, when it did appear, it was expurgated. A whole eighty-six-page section on CIA plots to assassinate foreign leaders was omitted.[17]

When it became clear that there had been CIA transgressions under the Nixon administration, Republican politicians seized on a counter-vailing opportunity. Instead of arguing against transparency, they demanded more of it. They aimed to dilute the iniquities of Republican administrations by placing them in the same mix as the transgressions of Democratic presidents. Former Republican presidential candidate Senator Barry Goldwater added a gloss when he insisted that presidents, Democratic as well as Republican, were more responsible for the CIA's transgressions than the agency itself.

Democratic presidential icons from FDR to JFK were in danger of being discredited. Church was determined to protect the reputation of John F. Kennedy, but Goldwater complained "All we've been hearing is pro-Kennedy witnesses." To escape this situation, Senator Church mooted a new idea. President Kennedy's former national security adviser McGeorge Bundy had suggested to him over breakfast one day that the CIA acted like "a rogue elephant on the rampage." It was an attractive image politically, for it exonerated Republican and Democratic presidents alike, and to propose such a thing had the ring of statesman-ship in the interest of national unity at a time of bitter divisions. On 18 July, Church announced at a press briefing, "The agency may have been behaving like a rogue elephant on the rampage."[18]

Later historians are agreed that, while the White House directed the CIA most of the time, the agency did in some instances act in a maverick fashion. For example, when President Nixon engaged in international treaty negotiations aimed at the harmonization of weaponry, he ordered the destruction of all biological and chemical warfare supplies, but a CIA officer disobeyed and held on to the agency's store of shellfish toxins and the dart guns with which to deliver them.

At the time, though, Church's rogue elephant proposition met with a hostile reception. Daniel Schorr was not alone in thinking that the rogue elephant remark showed that Church was "nursing presidential dreams." Church did have presidential potential. He had certain disadvantages, for example, he came from a small state and had poor health, but he was a man of clear abilities and was an able orator. He did finally declare his candidacy in March 1976. His defenders insist that he did not use the intelligence investigation to promote his presidential bid. They maintain that, on the contrary, the time- and energy-consuming inquiry eroded his chances of winning the Democratic nomination and contributed to the failure of his bid. Today's debate over Church's character and intentions is evidence of the durability of the 1970s political divisions over the CIA.[19]

In May 1975 David Atlee Phillips took early retirement from a senior position in the CIA to promote a fightback against defamation and reform. He organized 600 fellow spy veterans into the Association of Retired Intelligence Officers (ARIO). The CIA was in no position to be overtly political in its own defense, so Phillips and his colleagues stepped into the breach, writing op-eds, training speakers to visit Rotary Clubs and the like, and preparing testimony in defense of the agency. Phillips insisted that a speaker defending the CIA "must be *principled.*"[20] He conceded that there had been errors in the past and mounted sophisticated defenses. He opined that labor strikes and inflation had ended Chilean democracy, not the CIA, but refrained from Cold War rhetoric: "[President Salvador] Allende was a Marxist dedicated to carrying out his revolution within a framework of the constitution—a laudable goal unprecedented in history." The CIA did carry out covert operations, but Philips made the discerning point that their impact paled by comparison with "our entire foreign policy, our foreign aid program, and our tariff policies." Phillips, Ray Cline, and their ARIO colleagues were the Jesuits of the pro-CIA counter-reformation.[21]

One of the ARIO's themes was that constant revelations posed a threat to the lives of CIA officers and agents operating abroad. Philip

Agee's actions seemed to prove the point. A CIA apostate and convert to Marxism, Agee was a prominent example of those who thought the best way to stop the machinations of the CIA was to expose its agents. The agency knew he was disaffected. It had sent a female agent to Paris to keep him under surveillance and control his behavior. She bugged his typewriter.[22] But Agee saw through all that, and in April 1975 the UK's Penguin publishers brought out his book *Inside the Company*, which listed all the agents and front organizations Agee could remember from his time as a CIA officer in Latin America.

Matters came to a head on 23 December 1975, when four unknown assailants shot dead the CIA's main man in Greece, Richard Welch, outside his residence in Athens. It later transpired that the gunmen belonged to "17 November," a terrorist group that wanted to overthrow the Greek government in spite of the recent restoration of democracy in that country. It also turned out that the unfortunate Welch had taken no security precautions and that his address was well known. But, for the priests of the pro-CIA counter-revolution, this latest assassination was a golden opportunity to argue that enough was enough, that naming names was tantamount to treasonous murder, and that the campaign against the CIA had gone too far.

Within weeks, another scandal undermined the CIA's critics. The Pike Committee had prepared a report that was critical of the CIA's covert operations, listed its intelligence failures, and mounted an attack on Henry Kissinger in a section headed "SALT—Political Control of Intelligence."[23] In light of its sensitive subject matter, the House of Representatives voted to withhold the report from the general public. However, on 11 February 1976 New York City's *Village Voice* magazine published an unauthorized version of the report that they had obtained from Daniel Schorr, who was privy to the Pike Committee's work. Otis Pike and his colleagues now came under attack, unjustified but damaging nevertheless, for having leaked their own report.

The agency's critics were in disarray, and President Ford seized on the moment. All of a sudden, Colby was a man who talked too much. In the fall of 1975, Ford replaced him with a director, George H. W. Bush, who was committed to discretion and to the restoration of morale in the shaken agency. Ford now announced a reform plan that fell short of the expectations of congressional reformers. As he saw it, the CIA needed to be strengthened, not constrained. If its surveillance activities had been illegal, they should not be stopped but given a legal framework, and here the president appealed to Congress for help.

Additionally, Ford issued an executive order prohibiting the use of assassination. His motive in doing so may have been preemptive, an attempt to substitute executive for legislative action. Ford's action meant that assassination had been banned but not made illegal. One of Ford's speech drafts stated, "We must never bind our own hands so tightly that we become a helpless giant in a very real and very hostile world."[24]

A rejuvenated enthusiasm for secrecy held the White House in its grip. The affair of the *Hughes Glomar Explorer* was influential here. When it discovered the location, in the Pacific Ocean, of a sunken Soviet submarine, the CIA had decided to lift it to examine its military and encryption technology, and to that end paid the reclusive billionaire Howard R. Hughes $350 million for the use of his deep-sea prospecting vessel *Glomar Explorer*. Modified to give it lifting capability, the *Glomar Explorer* made an abortive attempt to raise the submarine in 1974. Farcically, the operation, to which hundreds of contractor personnel were privy, was supposed to be top secret but must have been obvious to Soviet intelligence. To save face and preserve the vestiges of détente, Moscow colluded with Washington in a conspiracy of silence. The taxpaying US public remained, however, in the dark. Some details of the operation leaked to the press in 1975, a year that was full of intelligence scandal, but the Ford administration as well as its successor kept pressing on the lid through which the genie had already escaped. In response to an inquiry about the operation in 2016, the former director of the *Glomar* project wrote, "We can neither confirm nor deny the existence of the information requested, but hypothetically, if such data were to exist, the subject would be classified and could not be disclosed."[25] His language reflected the post-1970s formula for negative responses to those seeking greater transparency in the conduct of national security policy.

The counter-reformation against intelligence reform was a powerful movement with real repercussions. Church and his allies nevertheless succeeded in certain areas. Both houses of Congress established permanent select committees on intelligence, committees that would play a leading role for decades to come. In 1978, with the Democrat Jimmy Carter now in the White House, Congress passed the Foreign

Intelligence Surveillance Act. As President Ford had desired, this legalized certain domestic activities such as bugging and wiretapping to fight foreign espionage and the bourgeoning threat of terrorism. However, under the new surveillance law, permission to bug had to be sought from a team of specially qualified judges, safeguarding civil liberties.

Finally, in 1980 the Intelligence Oversight Act made it a legal requirement that the CIA should report its proposed covert operations in advance to a select group, the "gang of eight," drawn from the intelligence committees. This briefly worded law gave force to the rump of a broader CIA charter that the Church Committee had drawn up, but which had suffered the checkered fate of the demands of Martin Luther's "ninety-five theses" four and a half centuries previously. According to Loch Johnson, who had served as a staff member on the Church Inquiry, the Oversight Act was "truly a striking step forward in the establishment of meaningful intelligence accountability— assuming the executive branch would honor the statute's wording and penumbra of intent."[26]

Like the Reformation of the sixteenth century and like more recent events such as Radical Reconstruction after the Civil War and the New Deal of the 1930s, the movement to reform the CIA in the 1970s stimulated not just a political reaction but also long-lasting controversy. One historian has contended that there had been adequate congressional oversight of the CIA ever since its inception.[27] This raises the question: if so, why was there an outburst of hostility toward the agency in the 1970s? As we have seen, the attacks on the CIA in the 1970s happened not just because of supervisory shortcomings but also because of other powerful factors such as détente and Watergate.

A more prevalent line of attack is that the reforms of the 1970s achieved very little of lasting importance. One explanation advanced to explain this is that the permanent committees set up to supervise the CIA suffered from the effects of that very permanence: official interactions between CIA leaders and their supervisors became social interactions, interactions that moved from coffee to alcohol, and to friendships whose warmth was incompatible with hostile questioning.[28]

Executive resistance to reform was a factor here. President Ford stuck to a recipe of mild changes and unrepentant operations. For Secretary of State Kissinger initiated Cold War-style covert operations under the very noses of Church, Pike, and their ilk. In Angola, where the

Portuguese were raping the country one last time as they relinquished colonial power, the CIA poured assistance into a faction aligned with apartheid-ridden South Africa, until Congress stepped in to kill the operation.

Even in a predominantly white nation like Australia there were accusations of CIA interference. There, the Labor prime minister Gough Whitlam challenged the activities of his country's CIA-aligned intelligence agency and talked of asking the United States to close down the signals intelligence listening posts it had established in the north of the country in accordance with the Five Eyes agreement. At this stage, using the royal prerogative and cheered on from the sidelines by Prince Charles, the heir to the British and Commonwealth throne, the governor general of Australia, Sir John Kerr, undemocratically removed Whitlam from office.

As in the case of Allende, the stated rationale was that Whitlam was ruining the economy. However, Whitlam had his supporters and feeling against the CIA ran high in Australia. Conspiracy theories began to flourish to the effect that the agency had engineered a coup in its fellow "Anglo-Saxon" former British colony. Those resentments did not subside when Australians learned that US Assistant Secretary of State for East Asia Warren Christopher promised the deposed Whitlam that "the US administration would never again interfere in the domestic processes of Australia." Although suggestive, the remark was inconclusive as proof that interference had actually taken place. But so many proven CIA interventions had occurred elsewhere that anti-CIA feeling could flame in the hearts of some Australians even in the absence of real fuel.[29]

The purpose of the Church reforms had been to change the CIA in a way that restored its standing and its fitness to protect the nation. In the United States, the wisdom and efficacy of the reforms were the subject of intense and long-term debate. However, in several foreign nations there was a belief that reform had taken place and that the United States had created an example to be followed. Canada, Australia, the United Kingdom, Denmark, and the European Union were amongst those who emulated America's attempt at "government in the sunshine." The perception that supervisory powers were in place also encouraged influential citizens of foreign lands openly to admire the CIA for many years to come. When his nation looked like being overwhelmed by international terrorism in 2015, Belgium's prime minister Charles Michel exclaimed in the words of one headline, "Il faut [créer] une grande CIA européenne" ("We must have a great European CIA"). And, of crucial importance, the perception that the CIA was now sensibly restrained potentially contributed to the greatest goal of all, the mutual trust achieved by Washington and Moscow in the 1980s.[30]

9

The Collapse of Soviet
Communism in the 1980s

The history of the CIA in the 1980s is inseparable from the demise of communism in Europe and the end of the Cold War. The roster of events is definitive. In 1982 President Ronald Reagan announced his successful drive for a Strategic Arms Reduction Treaty (START), which, by its final implementation in 2001, would achieve an 80 percent reduction in the number of nuclear warheads held by the US and the USSR. The Intermediate-Range Nuclear Forces Treaty of 1987 confirmed the progress being made and heralded the end, for the time being at least, of the arms race. Two years later, Soviet troops withdrew from Afghanistan, where they had been holding the line against CIA-supported Muslim fundamentalists, the mujahidin ("strugglers"). There were parallel Red Army withdrawals from Hungary and Mongolia. In November of 1989, joyous citizens dismantled the Berlin Wall, which had protected the communist east of the city from Western ideas and visitors since its erection at the height of the Cold War in 1961. In the following month, General Secretary Mikhail Gorbachev met President George H. W. Bush on the island of Malta. The Soviet and American leaders declared the Cold War to be over.

At home, Gorbachev was introducing reforms such as a new tolerance of private enterprise, and former member states of the Soviet Union were breaking free from Russian control. It spelled the end of communism in Europe. The true beneficiaries and actors in these events were the citizens of the former communist states. The events, as European citizens thought then and as European historians have said since, were episodes in their own continent's history. Typically, European historians hardly spare a word for the United States, let alone the CIA, when they explain what happened in the 1980s.[1]

But then, historians in general tend to ignore the role of intelligence. Often they are justified in doing so, but sometimes they are not. In the case of the demise of European communism and the end of the Cold War, claims have been made for the role of the CIA, and these invite consideration.

One interpretation that affords a place to the CIA is the victory hypothesis. According to this hypothesis, the CIA penetrated the darkness beyond the Iron Curtain and rightly estimated that Moscow was intensifying its military preparations. Armed with that information, the United States escalated its own defenses, notably by planning the Strategic Defense Initiative (SDI). Popularly known as "Star Wars" after the 1977 movie of that title, SDI was a space- and laser-based system for destroying incoming enemy missiles before they hit their targets.

Reagan later recalled, "I've had to tell the Soviet leaders a hundred times that the SDI was not a bargaining chip." The CIA's Russian specialist Robert Gates put a different gloss on Star Wars:

Reagan was convinced that the United States, and the West more generally, could bring serious economic pressure to bear on the Soviet Union that would adversely affect their ability to keep the system going while maintaining their ambitious military programs.

Gates added, "American advocates of SDI have contended that it was this program that broke the back of the Soviet Union and contributed critically to its ultimate demise."[2]

Presented with SDI, Moscow did consider the option of further escalation in response. But the Soviet economy, while it had seemed to perform well in the years of industrial takeoff, was now stagnating. According to a later in-house encapsulation, CIA estimates showed "Soviet GNP rising from a little more than a third of US GNP in 1950 to a little less than half in 1965 to 60 percent in 1975 before falling to less than a half in 1990." The victory theory holds that the Soviet Union proved to be unequal to the challenge of matching US spending. CIA reports indicated again and again that the Soviet leadership was acutely opposed to and concerned about the Strategic Defense Initiative. So, according to the victory thesis, Gorbachev very sensibly threw in the towel. He agreed to arms limitation, free elections in satellite nations, and greater openness in Soviet society.[3]

Other interpretations delve into the shadowy world of deception. There is the possibility that the CIA exaggerated the Soviet military capability in order to help Reagan persuade Congress to swallow the pill of Star Wars' multibillion-dollar expenditure, with the express intention of bringing the antagonist's economy to its knees.[4] This otherwise persuasive argument needs to be qualified in two ways. First, as Undersecretary of State Kenneth W. Dam observed, Congress would refuse to fund Star Wars "unless we are shown to be willing to deal with the Soviets on arms control."[5] Second, according to Gorbachev's account, he embarked on reform because of demands and pressures within the Soviet Union, not because of external threats.[6] But then, Gorbachev would say that, as no national leader likes to admit his hand has been forced by foreign pressures.

One should also keep in mind the ancillary theory that the United States won the Cold War but lost the spy war. According to this theory, because the Soviet KGB penetrated the CIA, it was able to feed disinformation across the Atlantic, making Soviet military might seem greater than it was. By flexing its fiberless muscles in this way, Moscow may have played into the hands of Reagan's advisers, who wanted Congress to strengthen America's military posture.[7]

Much of the foregoing speculation rests on the assumption that intelligence had been politicized. Intelligence purists both inside and outside the CIA would throw up their hands in horror at the thought. A more pragmatic inquirer might ask: if such manipulation existed, did the said politicization of intelligence serve a useful purpose, and, if so, how effectively?

The Reagan administration (1981–9) did not invent the politicization of intelligence. It had been evident in estimates of Viet Cong strength and when Kissinger negotiated SALT I. It became an entrenched habit at the time of the intelligence flap, when Gerald Ford was president. When the Carter administration was negotiating SALT II, which aimed at overall stabilization of weapons and their delivery systems in anticipation of eventual reductions, conservative critics acted on the premise that there should be safeguards against politicization and in making that demand politicized the issue themselves. Kissinger was still behind the negotiations, which rendered them suspect in the eyes of conservatives who did not wish to be bamboozled twice by his trickery. President Ford and his appointee at the CIA, George H. W. Bush, agreed to the principle of "competitive estimating": a Team B, consisting of experts such as Harvard's historian of Russia Richard Pipes, would be given access to the same confidential data as Team A, the CIA's regular analysts.

Pipes was wary of the Kremlin, which in his view had "brilliantly exploited the political innocence of American scientists." Faced with competition from Pipes and his B Team, the CIA began to find reasons why it might be unwise to trust the Soviets. It found that the enemy spent not 5–7 percent of GNP on defense, as previously stated, but 13 percent. Perhaps in an effort to boost its hardline credentials, the CIA now accepted some of Moscow's too high estimates of Soviet economic performance—estimates previously dismissed as propaganda. When Team B-influenced assessments arrived on Capitol Hill, the Senate refused to ratify the agreement that President Carter had negotiated in good faith and signed in January 1979. In the event both sides to the SALT II deal agreed to abide by its terms even without ratification, making it a building block for the peace that broke out in the following decade. Yet the deliberations that preceded this implementation had been steeped in politicization and had cast a shadow over the CIA's reliability.[8]

The prospects for improved relations with Moscow did not look good at the beginning of the Reagan presidency. Secretary of State George Shultz was willing to make the effort, but was at odds with hawks in the cabinet such as Secretary of Defense Caspar Weinberger and William J. Casey, head of the CIA 1981–7.[9] Casey was an admirer of Donovan of the OSS and rebuilt his agency's covert operational capability, which had been run down under President Carter. He believed in taking the fight to the communists on all fronts. His elevation to the policymaking cabinet was a mark of the president's regard for him and signified a departure from the notion that the CIA should be removed from politics. As for the president, he spoke of the Soviet Union as an "evil empire" and promised to "unleash" the CIA. He thus burnished a reputation for anti-communism that already shone brassy bright.[10]

On 23 March 1983 President Reagan announced the Strategic Defense Initiative. Following a period of research and development estimated to cost $2 trillion, the United States would deploy laser-shooting, missile-busting satellites that would end forever the American people's exposure to nuclear holocaust. The policy shift made the Soviets nervous. When US-led North Atlantic Treaty Organization forces ran a nuclear attack simulation exercise called Able Archer in the fall of 1983, jumpy Soviet forces started loading atomic warheads onto their delivery systems. The obliteration of mankind was averted only at the last minute, after Soviet intelligence had helped to confirm that no attack was taking place. In the absence of US intelligence to explain Moscow's overreaction, the Soviets at first sight looked to be paranoid. The impression was confirmed when, in the following year, a Sikh nationalist assassinated Indian premier Indira Gandhi and Radio Moscow accused the CIA of being behind the atrocity.[11]

Moscow's fears appear in a different light if one takes into consideration the deployment, at the time of the Able Archer exercise, of a new generation of US cruise and Pershing-2 nuclear missiles. The Kremlin reckoned that they could obliterate Moscow within as little as six minutes of launch, which made these weapons far more dangerous than the Jupiters that had sparked the 1962 missile crisis. The superior American technology opened the possibility that a first strike would succeed, with the USSR being disabled before it could launch what could no longer be called its deterrent warheads.

Counselling caution in March 1984 about a further projected US-led "worldwide procedural nuclear command post exercise," a national intelligence council memorandum suggested that Secretary of Defense Weinberger should be made aware that, in light of Able Archer, "The Soviet level of concern may be considerably higher than generally believed."[12] In due course President Reagan came to realize why the Kremlin was so nervous. The Able Archer near miss had been a sobering experience for all concerned. Star Wars had been intended to keep the peace by protecting America, but in combination with a new generation of missiles it was a threat.

One might surmise that the Able Archer narrow squeak contributed to a mellowing in Ronald Reagan that led to renewed US disarmament overtures to the Soviet Union. We should note that there were those who questioned whether Reagan had a good enough grasp of intelligence estimates to play a significant role. Mythology held that the "great communicator" was too lazy and unfocused to read documents put in front of him. Norman Mailer claimed that Reagan was "the most ignorant president we ever had," and Gore Vidal once jested that the Reagan Library had burned down and "both books were lost," one of which Reagan was in process of coloring in. However, senior members of Reagan's administration denied this characterization, recalling that he read both daily briefings and longer-term national intelligence estimates. It seems that Reagan was a politician to his fingertips, knowing how to project an easygoing image while focusing on his mission behind the drapes of the White House.[13]

In agreeing in 1984 to fund $26 billion start-up costs for Star Wars, did Congress, perhaps like the president himself, act in response to twisted intelligence regarding a Soviet arms buildup? Deputy Director for Intelligence Robert M. Gates was the CIA's leading Soviet expert and would face distortion charges when, in 1991, the US Senate held hearings on his confirmation as director of the CIA. The controversial hearings would turn into the biggest inquiry into US intelligence since the congressional investigations of the mid-1970s. Gates later wrote that the charge of Soviet intelligence distortion was the issue "that truly imperiled my confirmation."[14]

One striking indictment of Gates came in the form of a written statement by Jennifer Lynn Gaudemans. In 1982, when studying for her master's at Princeton, Gaudemans had joined the CIA's Office of Soviet Analysis, known as SOVA, on a graduate fellowship. She had then worked for the unit until she left, disillusioned, in 1989. In her statement for the Senate Intelligence Committee, she accused Gates of removing personnel who did not agree with him, of imposing contrived terminology on CIA reports to please his political bosses, of obsessively seeing Soviet conspiracies around every corner, and of "blatantly pandering to one ideological viewpoint."[15]

More weighty opposition to Gates's confirmation came from one of his senior colleagues. Harold P. Ford was an experienced analyst who had at one point been chief of staff at the Office of National Estimates, by now renamed the National Intelligence Council, nicknamed Nick. Ford accused Gates of "skewing" intelligence and in particular of offering a "swerved vision" of the Soviet threat. He had been unable to predict "the probable fortunes of the USSR and the Soviet European bloc." He had delivered intelligence to please the hawks in President Reagan's entourage. This was unacceptable because a CIA director, Ford innocently insisted, should "independently and fiercely stand his ground with his boss, the President of the United States."[16]

Gates was certainly a single-minded individual. A Boy Scout in his youth, he had studied Russian history without visiting Russia and concluded that it was dominated by "oriental despotism." He dismissed détente as "born in Europe." Though he chided the Defense Intelligence

Agency for exaggerating the Soviet threat, he was never one to minimize it.

Gates supported predetermined policymaking tendencies. For example, twelve days before Reagan's announcement of Star Wars, he circulated a memorandum claiming that the Soviets were launching their own strategic defense initiative. In one simplistic or subservient assessment reliably attributed to him, Gates noted that, "With the traditional paranoia of a police state society...Soviet intelligence...overestimates both their vulnerability and our intentions."[17]

Gates remained at heart a Boy Scout glued to the flag. He upheld hard-line policy toward the USSR generally, and he clicked his heels in agreement whenever the White House upped the ante. Yet he did realize that Reagan was on a peace mission. When attending a September 1984 meeting, he slipped a note to a State Department colleague, "saying that the President was out in front of all his advisers" in his willingness to court the Soviet leadership.[18] Gates was a shrewd tactician who knew the right moves in the bear pit of Washington politics, even if he had limited international vision and could not see beyond apparent Soviet paranoia.

The CIA engaged in direct efforts to weaken the Soviet Union. Casey implemented policies that extended beyond the agency's normal range. There was, for example, the plan to sabotage the finances behind the proposed natural gas pipeline running from the Soviet Union to West Germany and to other democratic European countries as well. The costly pipeline would have answered some of West Germany's energy requirements, but would have made America's ally dependent on Moscow, and would have given the Soviets a much-needed foreign currency boost. More orthodoxly, the CIA used contacts available to Israeli intelligence to establish a relationship with Solidarity, the Polish resistance movement. The Poles needed no encouragement in their determination to throw off the shackles of communism, but the proffered help was a signal of CIA intent.[19]

If there seems to be a certain braggadocio in some of the claims about CIA actions against the Soviet Union, it is in part because its successes had to remain secret, details emerging only years later. For example, the then unknown Colonel Ryszard Kuklinksi, smuggled out of Poland in December 1981 by the CIA, was a source of confidential information on his country's politics and on Soviet military dispositions.

US covert operational efficacy nevertheless suffered setbacks in the 1980s. In 1985, it was revealed that Edward Lee Howard, a CIA officer fired in 1983 for drug indiscretions, might have been a KGB asset (though there were also suspicions that this was a Soviet bluff). Though it was not realized at the time, in the same year Aldrich Ames became a Soviet spy—he worked in the sensitive counterintelligence section of the CIA. The year 1985 could be represented as the time when the United States lost what the CIA referred to as the "spy wars."[20] For it was also in that year that the FBI's Robert Hanssen, after a break of a few years, resumed his spying activities on behalf of the Soviet Union. Perhaps a decline in Angleton-era paranoia had allowed the defections to go unnoticed. In a delicious twist of irony, that very diminution of suspiciousness may also have underpinned the leap of trust that led to the end of the Cold War.

The setbacks were undiscovered at the time and did not discourage the CIA's Kremlinologists. These researchers and analysts followed as closely as they could the progression of changes in the Soviet leadership. One after another, the Kremlin's top men succumbed to old age and illness. The ailing Leonid Brezhnev, who had been in charge since the departure of Khrushchev in 1964, died in 1982. His successor Yuri Andropov had a hard-line reputation, having contributed to the crushing of the Hungarian uprising of 1956 and the Soviet invasion of Afghanistan in 1979, but was, according to the CIA's estimate, weak on economic policy. In 1984 Andropov died of renal problems. Konstantin Chernenko then assumed the general secretaryship of the Communist Party in his place. A heavy smoker and already chronically sick, he died in March 1985. His demise created the opportunity for Mikhail Gorbachev, a younger man with ambitions for change who served as general secretary until 1991. The Kremlinologists at first assessed him as "aggressive" and "activist." However, Gorbachev introduced the ideas of *glasnost* (transparency) and *perestroika* (restructuring) and set out to end the corrupt old ways of Soviet communism.[21]

As Washington–Moscow negotiations opened and proceeded, CIA-generated data and appraisals were constantly at hand. In September 1985 the agency prepared a paper on the Soviet posture toward a Reagan–Gorbachev summit meeting planned for Geneva in November of that year. It warned of an impending Soviet propaganda campaign for the "non-militarization of space" that would be aimed at some of America's allies, but suggested Moscow might make "substantive"

concessions, were the US to abandon Star Wars. In a National Security
Council meeting later that month, Secretary of State Shultz promoted
the CIA report as "very useful." Economic difficulties could well force
the Soviets "to seek resolution of some of their international difficulties. We need to treat this possibility seriously and decide how best we
can take advantage of possible opportunities."[22]

As the days of the historic meeting approached, the CIA remained
engaged in the dialog. As usual, there was interdepartmental sniping.
Director Casey seemed sometimes more aligned with Defense
Secretary Weinberger than with his own analysts. Shultz could not rely
on members of his own department to support the conciliatory line of
the CIA's Soviet specialists. State Department intelligence—INR—
alluded on 14 November 1985 to a "sensitive CIA report that *purports*
to describe Gorbachev's goals." But the CIA people stuck to their view
that the Kremlin would open its door to US negotiators. In a draft
national intelligence estimate on the eve of the Geneva meeting, they
said that Gorbachev was making progress in undoing the damage of
the Brezhnev years, but would still not meet his economic targets. For
economic reasons and because of acute, nationalist discontent within
the Soviet Union, especially in Poland, Gorbachev would want an
accommodation with the United States.[23]

The meeting between Reagan and his Soviet counterpart duly took
place on 19–20 November 1985. On the 19th, the two men met with
their negotiating teams in an eighteenth-century Swiss lakeside man-
sion. Then they walked to a boathouse 100 yards away and sat in front
of a log fire, with only their translators present. They discussed disarmament. Reagan had decided to trust Gorbachev. As he put it:

That morning, as we shook hands and I looked into his smile, I sensed I had
been right and felt a surge of optimism that my plan might work.

In his autobiography *An American Life*, the president gave that encounter pride of place, making it the book's opening passage. It was an
acknowledgment that trusting Gorbachev was his greatest achievement. Reagan, a former B-movie actor who was never going to go
down well with the intellectuals, attracted scorn. One British reviewer
derided his "self-important accounts" and scorned his book as "swiftly
concocted by busy little Californian ghosts." Like other presidents,
Reagan did employ shadow writers, in this case Robert Lindsey. The
greater truth is that both the United States and the Soviet Union had

produced leaders with the courage to face down the facile nationalists in their respective countries.[24]

As Jennifer Gaudemans had noted, the findings of her colleagues were often ignored. SOVA director Melvin Goodman, who saw Gorbachev as a credible reformer, was regularly snubbed. Gaudemans and Goodman were not alone in their anger. In August 1986, with another summit meeting in the offing, the head of the CIA's arms control unit, Doug George, wrote to Director Casey about the way in which US negotiators were ignoring his briefings. Reading an account of their meeting with Soviet officials, he "gave up in disgust." They were unprepared and an embarrassment. The situation arose, he thought, because of poor policy direction and "personality conflicts."[25]

But someone was listening. When Reagan met Gorbachev in Reykjavik in October 1986, he surprised everyone by accelerating START and proposing a ban on nuclear weapons. Gorbachev at first dragged his feet, as previously demanding as a precondition an end to the Star Wars project. When the United States refused to accede, Gorbachev finally shifted his position. Star Wars might have been provocative, but it was a long way from deployment, and less threatening the more one looked at it. CIA analysts supported the president's willingness to trust an old enemy and take a risk, notably over the issue of Soviet cheating and concealment. They favored an agreement that would ban concealment that prevented verification, challenging the Department of Defense, which insisted on the harder line that there should be a ban on all concealment of weapons covered by a START agreement.[26]

In January 1987 William Casey, who was losing his fight against an aggressive brain tumor, stepped down from the directorship of the CIA. William H. Webster, who had been director of the FBI since 1978 and now succeeded Casey on 26 May 1987, brought an additional touch of flexibility to US–Soviet relations. The new CIA chief was no admirer of the Soviet first secretary: "Gorbachev in our judgment had many admirable qualities, but he was raised as a Communist and never stopped being a Communist."[27] But Webster appears to have taken seriously the emollient findings of his Soviet analysts. They warned that Gorbachev was coming under attack at home for his innovative policies, which gave some urgency to the need for an international détente that would benefit both parties.[28]

The upshot was that in December 1987 the rival powers signed an epochal arms limitation treaty, freezing the numbers of intermediate-range nuclear missiles and providing for verification inspections, a first for the hitherto closed Soviet Union. Following the agreement, Gorbachev allowed proper democratic elections to take place in the constituent republics of the Soviet Union. He prized open the clammy hand of the censors, and in 1988 Russian citizens were for the first time able to read *Doctor Zhivago* without fear of reprisal. In an address to the United Nations in December of the same year, the Soviet leader promised to withdraw 500,000 troops from Eastern Europe and to reduce military spending by 14 percent.

Although there is evidence to suggest that the CIA's analysts predicted internally the collapse of Soviet communism and the fall of the Berlin Wall, the agency did not issue a warning.[29] Such happenings were alien to the imagination of Robert Gates, whose whole career had been built on the idea of a constant Soviet threat that he would constantly report to his political bosses, and that would keep him permanently in employment. Critics added the predictive failures to their list of the CIA's shortcomings, and they played a part in the protracted hearings that ultimately led to Gates's confirmation as director of the CIA. Having survived the extensive grilling, Gates held the post from 1991 to 1993, a period when he was a confidant to a former CIA director, now president of the United States, George H. W. Bush.

Once the "victory" deed was done, the CIA does appear to have saved Gorbachev on one occasion. CIA technician Tony Mendez told the story in his 2019 book, *The Moscow Rules*. First, he impressed on his readers that Soviet counterintelligence was outstandingly good. By way of illustration, he recited the incident of a CIA officer who was approached on the Moscow to Leningrad express train by a man who gave him an envelope. The envelope contained a note announcing two things. First, the man was the CIA spy's KGB case officer. Second, he wanted to become a double agent. The CIA officer was excited for a short period, until he learned that lots of KGB guys were offering their services. It was a plot to confuse the CIA. If a real defector were to offer his services, how would the agency know if the offer was genuine? In other words, even after the CIA (according to Mendez) upped its game, the opposition remained formidable, and running secret CIA operations in Moscow remained very difficult.[30]

The preparation for the Gorbachev rescue operation began in 1989. In that year, when US–USSR negotiations were still under way, an American agent penetrated the maze of tunnels that had existed under the Kremlin complex since the days of Ivan the Terrible. Guided by a CIA mole who had entered the above-ground complex on the pretext of watching the ballet *Coppélia* in the Kremlin Theater, the agent bugged the KGB's communications system. In August 1991—Gorbachev was by this time president of the soon-to-be-defunct Soviet Union— the bug was still operational and gave timely warning of a coup attempt against the man who had battled to end Russian autocracy.[31]

The rapprochement between the United States and the Soviet Union could not have occurred without the positive personal chemistry between the two powers' leaders. As Webster said of Reagan's take on Gorbachev, he "actually began to like him."[32] However, regardless of the attitudes of Gates, Casey, and Webster, regardless of corrosive interdepartmental rivalries, and in spite of the Ames betrayal, the CIA's Soviet specialists were always at hand with their informed perspectives. Their presence was a safeguard against Moscow treachery and gave confidence to the nation's leader when he wanted to make a leap of trust.

10

Iran and Iran–Contra

In broad daylight on 5 October 1986, Eugene Hasenfus set forth to commit a crime. The ginger-haired ex-Marine and former CIA operative was a family man who until recently had resided with his wife and three children in Marinette, Wisconsin. He had given up his work in the Badger State's construction industry to savor once more the bucks and buzz of his old business.

Hasenfus hauled his large frame on board the C-123 transport plane. With him on board the prop-driven aircraft when it took off at 9:30 a.m. from Ilopango, El Salvador, were a young Nicaraguan tasked with maintaining air-to-ground radio control and two American pilots wearing heavy gold jewelry. The pilots hurled good-humored epithets at 45-year-old Hasenfus for being so chicken as to don a parachute at the commencement of the flight.

The plane droned its way down the Pacific coast and entered what its crew understood to be enemy territory. The strongly built Hasenfus prepared to do his job as a "kicker." On reaching a drop spot, his sturdy legs would thrust out parachute-equipped pallets containing supplies for the Contras, the military forces opposing the democratically elected left-wing government of Nicaragua. The intention was that by the time the C-123 turned west and north to return to base, it would have jettisoned its entire cargo of jungle boots, AK-47 automatic rifles, grenade launchers, and ammunition.

In the tropical jungle below was the Gaspar García Laviana Brigade, named for a revolutionary priest. Its patrols were on the lookout for foreign incursions. On most days, there was little prospect of any action. But today, the members of one patrol heard the approach of the C-123, flying low. Looking up, a young soldier caught sight of the looming aircraft. He raised his SA-7 Soviet-made surface-to-air missile launcher and fired. This was at 12:38 p.m. Moments later, the soldier leapt in jubilation to celebrate his hit. The plane entered a fatal descent. Soon a parachute emerged, below which dangled the sole survivor of the airborne gang, Eugene Hasenfus.

The whole operation had been amateurish. When the soldiers captured Hasenfus, he still carried identity papers. The wreckage of the aircraft yielded further documents pointing to the originators of its mission. In vain did America's agents in El Salvador try to conceal the remaining planes of the arms supply squadron by digging a large hole and burying them. When Hasenfus appeared in a press conference and revealed his CIA connection, the secret was out.[1]

The Hasenfus indiscretion primed the American media for another revelation just under a month later. The news story appeared on 3 November 1986 in Lebanon's venerable weekly news magazine, *Ash-Shiraa*. This Beirut publication told of a visit to Tehran, back in May, by Robert C. McFarlane, until recently President Reagan's national security adviser. The news was startling because Iran and the United States were supposed to be at loggerheads. The official Iranian line was that the United States was the "Great Satan," an immoral imperialist power. Washington's line was that Iran was a terrorist-promoting Islamic rogue state, a finding that justified the US arms embargo then in place. Yet McFarlane had arrived to do business. It was a closely guarded secret until the American press picked up on the *Ash-Shiraa* story.

The US backchannel emissary was in Tehran to propose a deal. Iran and its neighbor Iraq were at war—it was a brutal conflict that ran from 1980 to 1988 and resulted in more than 300,000 deaths. Equipped with US-made weaponry sold to the shah's regime back in the 1970s, Iran's military needed spare parts but was stymied by the arms embargo. McFarlane offered to supply the deficiency. In return, Iran was to use its influence to secure the release of US hostages held in Beirut, notably the CIA head of station in that city, William F. Buckley. Around one hundred such hostages had been kidnapped by Islamic Jihad and Hezbollah, terrorist groups that supposedly responded to orders from Tehran. According to *Ash-Shiraa*, McFarlane also demanded that Iran end its support for the "liberation movements of the world."[2]

The *Ash-Shiraa* story was the beginning of the revelation that became one of the twentieth century's great scandals, namely, that irresponsible individuals within the US governing establishment were not only supplying Iran with weapons but also diverting the profits to finance the Contras in Nicaragua, an activity that Congress had specifically outlawed.

The Iran—Contra scandal was the latest example of a long-running, disastrous strain in US foreign policy. This was Washington's opportunistic relationship with Iran stretching back to 1953, the year in which the CIA had played a part in terminating Iranian democracy with its allegedly wayward tendencies. In the fifties, the Eisenhower administration reasoned that, in the restored monarch or shah, it had acquired a strategic ally in a vital strategic area. Geographically, Iran was positioned to block Soviet expansion southward to the warm-water ports that it coveted.

Iran was doubly precious in its contiguity with the Soviet Union. At Behshahr and Kapkan, communities strung along the Alborz mountain range near the Soviet border, the CIA would establish listening posts conveniently close to the communist bloc's missile-testing sites. Also geographically, Iran occupied a controlling position on the oil-producing Persian Gulf and the Strait of Hormuz that gave access to it. Iran itself was oil-rich. In sum, Iran, together with Israel, comprised a building block of US power in the volatile Middle East.

The Eisenhower formula worked for a while. Yet, within Iran and beyond, it fomented hatred of America and ended in strategic defeats. In the deceptively stable years, the CIA remained ruinously involved in the affairs of America's apparently stable ally. Iranians who opposed the shah's monopoly of power saw the CIA as the sinister hand, second only to the British in perfidiousness, that guided their nation's destiny.[3]

In the 1950s the United States propped up the regime of Muhammad Reza Shah Pahlavi with mainly military aid. Officially granted to stiffen resistance to any Soviet incursions, in practice the lethal hardware empowered domestic repression. The shah used his power to squeeze and eventually crush the Tudeh, his country's communist party. The CIA helped by promoting a new Iranian domestic security police. Established in 1958, the SAVAK became a byword for repressive brutality. In vain did analysts within the CIA point to the reactionary nature of the absolute monarch and lament the absence of social reform.[4]

In 1963 the shah addressed that problem by launching his "White Revolution." His decrees obliged the great landowners and tribal leaders to give up tracts of their land in return for shares in the nation's industrial enterprises. The newly available land was parceled up and given to 2.5 million landless families. In addition to appealing to the peasantry in this way, the shah promised better conditions for urban workers, set up a literacy program, and allowed women to vote for the first time. In subsequent elections, including a plebiscite, the shah won approval for his changes. However, American observers noted that the elections were marked by rigging and intimidation. The shah called for better education but—one way of achieving sales—made his own book, *The White Revolution*, compulsory reading in all schools.[5]

The White Revolution ran into cultural obstacles. As other nations had already discovered, peasant families are not necessarily helped and appeased by the gift of small plots of land. Women who voted and wore lipstick might have seemed progressive to the Swiss-educated shah and his Western promoters, but they were anathema to Iran's ulamas, or religious leaders, who relished the opportunity to step into the space vacated by the Tudeh.

It would be gifting the CIA with too much foresight to say that its analysts foresaw the triumph of the ulamas, but in the 1960s those analysts did remain pessimistic about the shah's chances of survival. In spite of the analysts' concerns, it remained US policy to support the shah, who became even more indispensable when Israel fought its Arab neighbors in 1967, weakening for a while the Israeli pillar of Middle Eastern American diplomacy.[6]

The United States was supporting an unpopular ruler, and many Iranians continued to blame America for the shortcomings of their governing regime. In 1963, when the shah's reforms were at their pinnacle, a West German poll of young Iranians indicated that they believed America, not the Soviets, to be aggressive. Eighty-five percent of those polled thought that the United States conspired to "make the rich richer."[7]

According to a widely ventilated vein of criticism, the CIA, departing from its 1950s and 1960s pessimism about the shah's regime, became complacent by the 1970s. As the historian James Bill put it:

> The State Department and CIA analysts in Washington always worked in the shadow of "the Pahlavi premise," the powerful assumption in the highest echelons of the executive branch of government that the pro-American and anti-communist shah was in complete control of Iran.[8]

Specific strands of the indictment included the assertion that the CIA failed to switch from seeing communism as the main enemy and showed an inability to focus on Iran's emergent clerical radicalism. One ayatollah (high-ranking ulama) in particular slipped the net. This was Ruhollah Khomeini, who inspired resistance to the shah from his exile in Iraq and would return in triumph to take charge of Iran in February 1979. The CIA was also charged with an oversight in health intelligence. Unbeknownst to the Carter administration, the shah was dying of cancer. Faced with disturbances, in September 1978 the shah declared martial law, but his declining health made him ill-equipped to follow through when even more serious rioting broke out. By the time Khomeini arrived in his homeland, the last shah of the Pahlavi dynasty had fled.

As so often happened, the CIA served as a scapegoat for executive unawareness. To a degree, the opprobrium leveled at the agency was appropriate. Over the years, the CIA's intelligence specialists had been relegated to second place behind the operators charged with support-ing SAVAK. Its team of Iran analysts had been reduced to a rump. Yet in spite of these shortcomings the CIA did supply timely warnings— warnings that fell largely on deaf ears.[9] In May 1978, the CIA reported that the shah was "gambling" that his reforms would be sufficient to prevent the clerics from returning Iran to "the Middle Ages." In June, it warned of impending unrest and of the influence of the "politicized clergy."[10] In November, the agency warned that "the eloquent and charismatic Khomeini" held the Iranian people in his grip and that nobody dared oppose him. The CIA's chief analyst, Robert R. Bowie, urged that, to save the day, the shah would have to go.[11]

The CIA warnings did not alter the rigid stance of President Carter's national security adviser, Zbigniew Brzezinski. Polish-born and anti-Russian, Brzezinski was convinced that all threats emanated from the Kremlin. He paid scant heed to the CIA and failed to make the transi-tion required to understand Iranian clerical issues.[12]

Executive inattention to Iran had grave consequences. Brzezinski's attempts to blame CIA somnolence for the collapse of US regional strategy did little to save the Carter administration's reputation. On 4 November 1979 that reputation took an irreversible plunge when a group of Iranian students decided on direct action. Initially some of the students were left-wing, and some pro-democracy, but members of the Muslim Students Association took control. They reacted when they learned that the shah had been admitted to the United States. The 60-year-old fugitive ruler had traveled to New York City in the desperate hope that American doctors could stave off the effects of cancer and save his life. That circumstance had no effect on Iran's radical students. One of the embassy captive takers recalled how she and her fellow students feared a return to 1953: "In the back of everybody's mind hung the suspicion that with the admission of the shah to the United States, the countdown to another coup d'état had begun." Believing that they had the blessing of Khomeini—though the ayatollah does not seem to have issued an order—the students seized the US embassy in Tehran, taking hostage sixty-six personnel who were at the embassy on that day. They demanded that the shah be given over to their custody. There were doubts within the newly installed Iranian regime. But Khomeini, who became Supreme Leader of Iran on 3 December, declared his support for the embassy occupation, an occupation that violated the Vienna Convention on diplomatic immunity.[13]

It was a testing time for the Carter administration. On Christmas Eve the first contingent of 30,000 Soviet troops invaded Iran's neighbor Afghanistan in an effort to prop up its socialist regime. The CIA had been quietly undermining that regime and, once the Soviets arrived, it redoubled its efforts. The agency's encouragement and support of Afghan fundamentalists helped to ensure that Moscow's decision to invade Afghanistan was one of its more serious mistakes. There was a long-term downside to this, in that the United States in due course had to confront the very fundamentalists it had helped nurture. As for short-term benefits, they were not apparent at the time.[14] As of the last year of the Carter administration, the United States seemed to have lost influence in a vital region of the globe and appeared to be weak in the face of advances by the Soviet Union.

Earlier in his presidency, Carter had acted in a way that reflected mid-1970s concerns about real and alleged CIA excesses. In a move that provoked fury, he had reduced the covert operational capacity of the agency, firing hundreds of personnel. Now, like other presidents before and since, he changed his tune and looked to the operators for a quick fix. Could the CIA and military rescue the hostages? The CIA was reluctant to play ball. Admiral Stansfield Turner, who served as director of the CIA from 1977 to 1981, warned on 15 November 1979 "that any rescue operation into Tehran within the next two weeks would have a very low probability of success." He reminded Carter that when John F. Kennedy had authorized the Bay of Pigs invasion on the basis that it had a "fair" chance of success, Kennedy had failed to appreciate that "to the military, 'fair' is like a 'D' in college."[15]

Brzezinski was insistent on the need for an urgent response and appears to have wanted a clandestine rescue operation without approval in advance by Congress. Here, Turner had a further warning. After consulting with his agency's lawyers, he concluded that any rescue attempt involving the CIA would likely infringe the Hughes–Ryan Amendment of 1974, which had required advance notification of Congress of any contemplated covert operation.[16] Another source of anxiety was the preexisting bad reputation of the CIA in Iran. Initially, there was some consolation in the belief that "virtually all CIA documents in the Embassy were destroyed," which meant the full extent of the agency's past activities would never be exposed. But it emerged that the CIA's shredding machine in the embassy compound had broken down just a few minutes into the destruction operation. Other documents in the main embassy building had been successfully shredded. However, the Iranian students would reconstitute them. They published painstakingly reassembled documents in a collection that ran to fifty-eight volumes. Meanwhile, a member of the embassy staff publicly confessed to his CIA associations.[17]

On 27 January the CIA's undercover specialist Tony Mendez teamed up with Canadian operatives to execute a plan to rescue six further US foreign service personnel who had hidden in the private homes of Canadian officials. Armed with forged Canadian passports and given dummy identities as members of a fake movie company called Studio Six Productions, the Americans made a daring and successful dash for freedom.

A special Iran task force within the CIA now looked at options for freeing the much larger group detained in the embassy. What if Khomeini or the more pragmatic leadership of the Iranian government could be persuaded to rescue the detainees and hand them over? The agency estimated that the students would resist and there would be a bloodbath. Brzezinski turned to another type of unilateral action.[18]

Against the advice of Secretary of State Cyrus Vance, Carter reluctantly authorized a military-dominated rescue attempt that went ahead on 24 April 1980. The effort failed when a combination of mechanical failure and a dust storm disabled three of the eight helicopters involved. A remaining helicopter crashed in the subsequent retreat, killing eight Americans and an Iranian interpreter. Vance resigned, liberals in Congress were aghast that the operation had taken place, conservatives seethed at the failings of the liberal Carter administration, and yellow ribbons adorned millions of American homes in a display of support for the hostages.

The scene was set for the result of the 1980 presidential election, a defeat for Carter and a triumph for the Republican candidate, Ronald Reagan, on whose coattails the Republicans also seized control of the Senate for the first time since 1954. While the hostage issue was prominent in the election campaigns, it did not cause Carter's defeat. His administration had a dismal economic record. Interest rates and inflation were high, while productivity and employment levels were low. A majority of voters turned against what they saw as failing liberal nostrums and instead embraced the more conservative, business-oriented ideas of the Republican candidate.

The prominence of the hostage issue in the election campaign did, however, flag the intentions of the incoming president. Reagan badly wanted the hostages back in America. Iran's leaders had come to regard the detained Americans as a political embarrassment, and they did allow the embassy staff to return home—after 444 days of captivity. Their act of contrition did not occur, however, in time to ease the weight on President Carter's shoulders. The Iranians delayed the release so that it coincided with the day of President Reagan's inaugural address on 20 January 1981. Although Reagan's administration continued to label Iran a pariah state, the beneficial timing of the hostages' release proved to be a hint of things to come.

The Reagan administration's subsequent overtures to Iran arose from its policy toward a much smaller nation, Nicaragua. The people of that Central American country had every reason to feel antagonistic toward the United States. The US Marines had occupied Nicaragua between 1912 and 1933, and Washington had supported the Somoza dictatorship that followed. In the years 1974–9 President Anastasio Somoza Debayle was the latest of that pilfering dynasty. His dictatorship came to an abrupt end with the triumph of the Sandinista movement, named for Augusto César Sandino, who had led a rebellion against US occupation in the years 1927–33. In 1985, the Sandinistas, now in control, organized a presidential election. Their leader, Daniel Ortega, won handsomely. Perhaps more significantly for the durability of Nicaraguan democracy, when five years later Ortega lost an election to Violeta Chamorro, he would relinquish power peacefully—he would not assume power again until 2007, when, again legitimately, he won a further election. Ortega's 2007 victory would have sealed

Nicaragua's reputation for democratic legitimacy, but for his increasingly authoritarian tendencies since then.

Ortega was a socialist. Predictably, if improbably, the Reagan administration denounced him as a threat to democracy. On 1 December 1981 President Reagan authorized the CIA to fund covert operations against the Sandinista regime. A year later, Congress adopted the Boland Amendment, which prohibited such aid—it was named for its Democrat sponsor, Congressman Edward P. Boland of Massachusetts, who was chair of the House Intelligence Committee. Boland and his supporters were upset, in part, because the president and his CIA director, Bill Casey, had a cavalier attitude toward congressional oversight. Asked about the legislature's role, Casey growled, "The business of Congress is to stay the fuck out of my business."[19]

By 1984 America's devotees of transparency were beginning to think that the bad old days had returned. There were stories about the Contras, the Nicaraguan faction supported by the CIA, that suggested they resorted to ruthless tactics, including rape. When the news leaked that the CIA had mined Nicaragua's harbors, it was too much for Barry Goldwater, who was by now chair of the Senate Intelligence Committee. Although the CIA was becoming one of the darlings of American conservatives, and although the senator from Arizona normally championed its cause, Goldwater did not like being kept in the dark and publicly informed Casey that he was "pissed off."[20]

Walter Mondale, the Democratic candidate who opposed Reagan in the 1984 general election, was a seasoned critic of covert operations. House Intelligence Committee chairman Boland assisted Mondale's cause by securing the passage of a further amendment, this time banning the use of third parties to finance the Contras. Then, in the month of October, when campaigning was approaching its crescendo, a story broke that seemed custom-built to strengthen Mondale's campaign. The Contras disseminated a manual on guerrilla warfare with instructions on terror bombing and kidnapping—and on how to assassinate people on your own side and defame your enemies by accusing them of being the killers. The American author of the manual, the heavy-drinking and emotional John Kirkpatrick, had contravened Reagan's reiteration, in a 1981 executive order, of the nation's ban on assassination.

Nicaragua charged the CIA with sending assassination teams against its leadership. The agency at first responded with a letter to one of Boland's colleagues on the House Intelligence Committee with a "categorical" denial, insisting it was upholding "both the letter and the spirit" of Reagan's Executive Order 12333 banning the use of assassination. Following further uproar in the press, the CIA's director wrote to another member of the House Intelligence Committee, defending his agency's conduct in Caseyspeak:

I'd like you to look through the much publicized text of the FDN [*Fuerza Democrática Nicaragüense*/Contra] manual on psychological operations together with the code of conduct prepared in pocket size for every FDN soldier to carry with him at all times.... They were prepared by the political section of the FDN with the help of an advisor provided by the CIA. The code of conduct explains that the objective of the FDN is the development of a democratic and pluralistic government in Nicaragua [and] to establish social justice and human rights.... Two of the four [assassination] passages were deleted by the FDN....[The fourth] uses the word "neutralize" in dealing with the problem of removing local officials on occupying a town.

When Mondale made an election issue of the manual, Reagan insisted that Kirkpatrick was not a CIA man, but a contract employee.[21]

Mondale's campaign gibes would be no match for Reagan's declaration that affluence had returned. The popular Republican candidate warned that the Democrats' high-tax, high-spend policies would spoil the American dream. The president won a crushing election victory, and although the CIA had once again been no more than a subsidiary issue, it was again possible to draw the conclusion that the administration had a renewed license to engage in clandestine operations.

The CIA's adventures—and suspicions about them—became wilder. According to the investigative journalist Bob Woodward, in 1985 Casey attempted assassination by surrogacy. He arranged for the Saudi Arabian intelligence service to eliminate the clerical scholar and suspected terrorist Sheikh Mohammed Hussein Fadlallah, head of the Hezbollah organization. The car bomb intended for that purpose exploded in a Beirut suburb, killing eighty people. It missed Fadlallah, who died peacefully twenty-five years later. Asked about the plan, Reagan would later say he had not ordered it, adding, "Never would I sign anything that would authorize an assassination. I never have." By the time he uttered those words, the Iran–Contra affair had shredded the credibility of his administration.[22]

The Iran–Contra scandal stemmed from plotting undertaken on behalf of the Reagan administration, plotting that invited exposure because of its amateurish character. Just as Congressman Boland feared, the Reagan administration was evading scrutiny by using third parties to funnel aid to the Contras. Reagan and his vice president, George H. W. Bush, agreed the diversion in principle in 1984. In that year the CIA secured the services of Manucher Ghorbanifar, an Iranian businessman and (it turned out) confidence man as a broker between Israel and Iran.

In dealing with Iran, Reagan had the apparent advantage that he was simultaneously authorizing aid to the mujahidin, a coalition of Islamic fighters struggling to free Afghanistan from Soviet control. Although his motivation there was anti-communism, his stance potentially gave him credibility in the wider Muslim world. However, the words "CIA" and "USA" remained terms of contempt in Iran, for Washington was openly backing Iraq's Saddam Hussein in his bloody war with his northern neighbor. Against this unstable background, the Ghorbanifar-brokered deal went ahead. According to the deal, Iran was supposed to use its influence to free American hostages, Israel would deliver US weapons to Iran, and the profits from that sale would be diverted to pay for the Contras' efforts in Nicaragua. In a refinement of the plot, Casey arranged for an official of the National Security Council, Oliver North, to run the operation independently of the CIA, which would keep his backchannel deal safe, so he vainly hoped, from the prying eyes of congressional intelligence committees.

In the fall of 1986, the unmasking of the Hasenfus mission in Nicaragua and the *Ash-Shiraa* story about Iranian shenanigans ignited the Iran–Contra scandal. Attorney General Edwin Meese had to confirm that federal dollars were syphoned to the Contras. The president fired Oliver North. The latter's photogenic secretary Fawn Hall added fuel to the fire when she was observed removing potentially incriminating paperwork from his office.

As was the custom in the face of damaging revelations, Reagan established a presidential commission to investigate. In charge of this damage limitation effort he placed John Tower, a former Republican legislator who had sat on the Church Inquiry in the 1970s. Reporting in February 1987, the Tower Commission concluded that Casey had been at fault in not taking charge of the Contra operation and in failing to notify Congress. From Reagan's point of view, it was a convenient finding, especially as Casey had by this time resigned as CIA director on grounds of ill health—he finally succumbed to the ravages of cancer in May.

Meese had by this time triggered the appointment of an independent counsel to investigate possible criminality in the Iran–Contra venture. Judge Lawrence E. Walsh issued fourteen indictments against Reagan administration officials, and court cases followed. Colonel North and his secretary had become iconic media figures and basked in the approval of many conservatives. Nevertheless, in May 1989 North would be found guilty on three counts, for which he was fined and put on two years' probation. Walsh also found against Assistant National Security Adviser John Poindexter and Secretary of Defense Caspar Weinberger, but the prosecutions got entangled in a welter of appeals. On Christmas Eve 1992 President George H. W. Bush would pardon Weinberger and five other Iran–Contra transgressors. Walsh reacted by complaining that he had struggled in the face of a cover-up. He summed up the attitude of the former CIA director thus: "While President Bush made numerous public statements extolling his cooperation with the Independent Counsel's investigation, that, in fact, had not been the case."[23]

As Robert Gates observed, the CIA became a "political football" in the course of the Iran–Contra debates.[24] Though a joint congressional inquiry went no further than to conclude that the Reagan administration had been unethical, some of its members signed a minority report blaming recent misdeeds on the plethora of Democrat-inspired legislation on what the CIA could and could not do. The Republican Bush was similarly defiant. He justified his valedictory Christmas Eve amnesties by claiming that the convictions of those pardoned had stemmed from the "criminalization of policy differences."[25]

The Iran–Contra scam did not contribute to President Reagan's US national security objectives. The Contras failed to accomplish what more peaceful Nicaraguans finally achieved at the ballot box—the defeat of the Ortega government. Iran did little to free US hostages. William Buckley's remains now rest honorably in Arlington National Cemetery, but they arrived there only after his gruesomely tortured body was discovered on a road near Beirut airport. Iran remained a dangerous opponent of the United States in the Middle East. America would have to fall back on its special relationship with Israel, a nation that could be counted on to provoke anti-Western Islamic *fatwas* ("findings" that could lead to terrorist killings).

Yet the affair cannot be said to have undermined the standing of the CIA. In moving the operational fulcrum outside the agency, Casey's motive had been the avoidance of scrutiny, but his maneuver also had the effect of diverting blame from his agency to other parts of the federal bureaucracy. Additionally, the Iran—Contra scandal played out against a triumphal background. The Reagan—Gorbachev interchange culminated in the arms limitation agreement of January 1988. A year later, Soviet forces withdrew from Afghanistan. CIA-shipped Stinger ground-to-air missiles had helped by enabling the mujahidin to shoot down Soviet helicopters. Those who wished to hype the agency could argue that, in Afghanistan and elsewhere, the CIA had strained Moscow's resources and given a powerful shove to the overthrow of communism back in Eastern Europe. The fall of the Berlin Wall in 1989 was more icing on the cake.

As we saw in the last chapter, the CIA cannot really be credited with having won the Cold War. It did draw sustenance from the reputation of having done so, and the "victory" hypothesis gave it a shield against Iran—Contra barbs. In a perverse twist, it would be precisely its association with Cold War victory, not its ill repute over recent covert diversions, that would threaten its standing and even its existence in the 1990s.

II

Existential Issues, 1990–7

In view of the existential issues this chapter will discuss, it is worth keeping in mind that the CIA continued to operate throughout the 1990s and continued to supply assessments based on hard-won information that policymakers in Washington were at liberty to heed. In Langley, efforts were made to improve the quality of the agency's work. They are evidenced in the initiative of John McLaughlin, deputy director of intelligence 1997–2000, who set up a school of intelligence analysis named for Sherman Kent, the respected theorist who had headed the Office of National Estimates for fifteen years from 1952. The school reflected the mushroom growth in intelligence teaching on American campuses in response to events such as the Family Jewels revelations in the 1970s and Iran–Contra in the 1980s.[1] At the new CIA school, experienced analysts passed on their wisdom to the next generation.

As in every decade since the CIA's formation, the agency's history shadowed the history of the entire globe. No attempt to trace its activities can be comprehensive. Here, we shall consider two examples as illustrations of the nature of its work.

Iraq is our first example. The country had fought a war against Iran with US military assistance, but that did not satisfy its leader, Saddam Hussein, who was displaying chronic aggression symptoms. For him, it was not enough that he had tried to annex a province of Iran. Now, he advanced Iraq's claim to sovereignty over an equally oil-rich territory, the nation of Kuwait. Baghdad accused Kuwait City of slant-drilling to steal Iraqi oil, and the assertion of Iraqi sovereignty was also a way of canceling the national debt, for to finance its war against Iran, Iraq had borrowed heavily from Kuwait. Saddam now mobilized his country's army for an invasion.

Though the CIA had no means of penetrating the political establishment in Baghdad, it had the technology to monitor Hussein's armed forces. In May 1990, it warned President Bush about the possibility of Iraqi aggression toward its southern neighbor. The Kuwaiti ruling family tried to fall back on self-fulfilling optimism. It asked Egypt to mediate, accused Hussein of bluffing, and even put its troops on a lower level of alert. A CIA daily briefing on 25 July 1990 took a more realistic approach and issued a Cassandra warning, "Iraq is probably not bluffing." On 2 August, Iraqi forces duly invaded and occupied Kuwait.[2]

What would happen if the United States intervened? The CIA submitted a wide-ranging assessment of Hussein scenarios—a coordinated summary, as usual, of the agency's own investigations and of parallel work by other agencies. Langley warned that Hussein might, if threatened, play the "Israeli card," representing himself as the Arab nations' defender against US–Israeli aggression. Further, if he was defeated by an American-led coalition and allowed to keep his armed forces intact, it would create the destabilizing danger of "an Israeli military attack aimed at eliminating Iraq's missile, chemical, biological, and nuclear capabilities."[3]

In the event the president, with his allies and with United Nations' blessing, launched an attack on Iraq's military in Kuwait, forcing it to retreat to its home territory. George H. W. Bush stopped right there without toppling Hussein's regime, a decision that his son would revisit. Whatever one's interpretation of these events, the CIA played an integral and thoughtful role.

We can follow this macro example of the CIA's utility with a slightly more ambiguous micro example. H. K. Roy, or a person styling himself with that name, was a CIA officer in the field. To put it less mundanely, he was an American secret agent. He was a man of firm convictions. He hated and disregarded the rule against CIA agents carrying firearms. He equipped his wife with a "tactical shotgun" to keep her safe in the hotel room to which she was so often confined—the CIA divorce rate, he drily observed in his memoir, ran high.

In the 1980s Roy helped run Nicaraguan operation; then in the 1990s was sent to Bosnia at a time when ethnic and religious bigotry were tearing apart multiethnic, interdenominational, and secular Yugoslavia, the federal state that President Woodrow Wilson has helped set up as a solution to Balkan rivalries. The dissolution of Yugoslavia yielded six states—Bosnia and Herzegovina, Croatia, Macedonia, Montenegro, Serbia, and Slovenia. Muslim-majority Bosnia harbored a Serbian, thus Orthodox Catholic minority that rebelled with sectarian assistance from another former Yugoslav state, Serbia. The Serbian insurgents were well armed. They massacred men, women, and children in acts of ethnic cleansing that amounted to genocide. The United States tried to redress the balance. Together with Muslim radicals who felt indebted to Washington for its help against the Soviets in Afghanistan, it contributed to the "Croatian Pipeline," a conduit for arms from Turkey and Iran smuggled via Croatia into Muslim-held areas of Bosnia. Saudi Arabian money financed the arms operation, and US-built, black-painted C-130 Hercules aircraft featured in the operation.

H. K. Roy was present at the siege of Sarajevo and witnessed the savagery unleashed against the Bosnian Muslims. The United States eventually brought military force to bear and secured peace through the Dayton Accord of November 1995. Roy was critical of Bush for initially trying to hold Yugoslavia together against the CIA's advice. He thought that even when Washington had tardily accepted the inevitable triumph of local nationalisms, it had been too slow to prevent the killing and mistreatment of Muslims.[4]

The reports of local agents like Roy filtered their way up to the CIA hierarchy from all over the world. They presented policymakers with a kaleidoscopic picture and the need to make challenging decisions about how to act in light of the circumstances. The reports were not always in tune with realpolitik. Osama bin Laden, destined to be a scourge of the United States, visited Sarajevo in October 1994. Roy believed that it was US indifference to the fate of Muslim Bosnians there that turned the future terrorist mastermind against America. Roy was right about the onset of indifference but not fully appreciative of the reason, namely, that serious doubts were setting in back at CIA headquarters and in Washington about the wisdom of cooperating with Islamic radicals like bin Laden.[5]

As the foregoing examples show, the CIA was fully active in the nineties. Nevertheless, in that decade the agency experienced a crisis. For the first time since the 1940s, it became necessary to justify the need for the CIA. President Truman's reason for creating the agency had been his desire to combat Soviet assertiveness, and to fight what came to be known as the Cold War. According to that logic, the CIA was no longer needed now that the Cold War was over. Congress, for its part, had in 1947 approved the agency's creation for the dominant reason that its members wanted the nation never again to be caught unawares by a surprise attack. By 1990 the proponents of that rationale were thin on the ground. A third reason for central intelligence, though it was not popular in the 1940s, was the notion that a nation as powerful as the United States was bound to make enemies, so in both war and peace America required access to other nations' secrets in order to thrive and survive. Policymakers finally accepted that rationale in the new post-Cold War era, but only after a turbulent period.

With the Cold War over, some dogs felt they had permission to bark. In spite of the fact that the CIA had contributed in the collective mind's eye to the triumphant outbreak of Soviet–American peace, all of a sudden there spewed forth a spate of accusations associated with that dark and unadmitted hole in the American psyche, failure. The Clinton administration, with its emphases on prosperity, trade, the environment, and diversity presented the agency with new challenges without extending the shield of protection that could be expected in the Cold War years.[6]

Of course, the CIA was no stranger to criticism. It had been charged with failing to predict Moscow's acquisition of atomic capabilities and later the collapse of Soviet communism. It failed to predict the downfall of the Berlin Wall, and there was a roster of other charges. But at the Robert M. Gates nomination hearings, a recent chief of CIA counterterrorism went further and asked the existential question, "Is the CIA relevant in the modern world?"[7]

The CIA's crisis pinnacled in 1994, when the CIA's Aldrich Ames turned out to be a double agent. The Ames affair was not unique. Between the CIA's inception and the end of the century, there were 150 cases of Americans spying on their own country. There had been a dip in the numbers at the height of the Cold War, then a rising trend since the reform-minded 1970s. But the 1994 episode was much more than a digit in a trend. Ames's treason was demonstrably more harmful to the CIA than any earlier episode.[8]

Ames was an experienced CIA officer who worked in the Soviet, later Russian section of the Operations Directorate's counterespionage section. Alcohol and womanizing got the better of him. His divorce cost him $46,000, and his mistress Rosario plenty more, sending money to her family in Colombia and buying 500 pairs of shoes. In April 1985

Ames had agreed to work for the KGB, receiving for his treachery sums totaling $4.6 million. He worked in a particularly sensitive section of the CIA and was able to give his Moscow controllers the identities or clues to the identities of serving undercover agents. One of these was Oleg Gordievsky, the KGB's man in London, who had been working as a double agent for British intelligence. Following up on leads supplied by Ames, the Russians recalled the suspect to Moscow. KGB counterintelligence agents had already noticed works by the banned novelist Aleksander Solzhenitsyn in Gordievsky's London flat. When they started asking their recalled emissary questions, Gordievsky knew it was time to leave his family and his homeland.[9]

MI6 spirited Gordievsky out of Russia just in time. The agency's exfiltration operative showed a Mars bar to reveal his identity and deployed an escape car equipped with a child and a picnic. Gordievsky disguised himself as a tramp and waited in a hedge for the arrival of the car. The ruse succeeded in his case, but others were less fortunate. The pensioner Dmitri Polyakov had been a senior officer in the GRU, Moscow's military intelligence service, and an informant who, for example, alerted the CIA to the Sino-Soviet split. Ames fingered him. Ames also identified a man who had been a prize CIA asset, the scientist and general Adolf Tolkachev. By supplying details of the Soviets' Su-27 and MiG-29 fighter planes, Tolkachev had caused modifications to America's F-15 fighters and saved the US military "billions of dollars and up to five years of R&D time." He, Polyakov, and Gordievsky were among the ten or more CIA agents whom Ames betrayed.[10]

It had become clear that the KGB must have a mole in the CIA. The KGB itself gave the game away. Instead of letting suspected agents run so that they could be followed, the time-honored practice of competent counterintelligence agencies, the Soviet authorities moved to arrest. Seeking plaudits inside the Kremlin, KGB's insecure foreign intelligence chief Vladimir Kryuchkov rushed to apprehend and then execute the agents identified by Ames. His actions triggered heightened awareness and frustration in Langley.[11]

In spite of these warning signs, Ames went undetected until his extravagant lifestyle became too conspicuous to be ignored. So there were accusations about too comfortable clubbiness in sections of the CIA. James Woolsey, who had succeeded Gates as director in February 1993, warned that the division of operations should no longer be run as a "white male...fraternity." In a later interview, he revived the populist canard. According to the Oxford-and-Yale-educated Woolsey, obsolete practices had meant that "the DO [Department of Operations] folks are sort of Ivy League guys who are fluent in French or something.... In '93 we still had all of our wonderful Cold Warriors and these were people that won the Cold War ... speaking their Russian and Polish and so forth."[12]

Their time had come and gone, Woolsey suggested. But time was up for Woolsey too. His assurances that one could trust the Russians to observe arms agreements now sounded a discordant note. His decision not to single out CIA colleagues for punishment over the Ames affair was taken as further evidence that he was "soft." The Senate Intelligence Committee issued a scathing report, and Woolsey had to resign. His departure from the hitherto coveted director's suite on the seventh floor at Langley would be one of several in quick succession. The CIA had no fewer than five directors in the period 1991–7, a sign of discontent both inside and outside the agency.[13]

This was not all. In 1995 Senator Daniel P. Moynihan introduced a bill for the abolition of the CIA. Some successful working-class people rise by conforming and by molding themselves in the image of their "betters." Moynihan, who in the words of his biographer was educated in East Harlem under circumstances where "no one would have given a cent for his chances of climbing to the top," was different in being a born controversialist.[14] For example, in the liberal-Democratic ascendancy of the 1960s, the senator's unsympathetic remarks about African American family life had made him enemies.

His remarks also won him conservative friends, and President Nixon offered the New York Democrat federal posts. Such preferment culminated in 1973 in his appointment to be the US ambassador to India. It was there that he developed his view about the potentially toxic effect of CIA activities. The agency's obvious as daylight espionage in India and ill-informed efforts to foul local communists were causing Indians to develop a paranoid attitude toward the CIA and, by extension, the United States. India's prime minister Mrs. Indira Gandhi, only the second woman in history to be elected a nation's leader and a major personality on the world stage, was appalled by CIA operations in Chile and feared the agency would dispose of her too. Moynihan appealed to Secretary of State and National Security Adviser Henry Kissinger to "pull the CIA out of India. It is a devastating liability."[15]

In the Reagan presidency, Moynihan was the minority leader on the Senate Intelligence Committee. When the majority leader, Senator Goldwater, said he was "pissed off" at the CIA's mining of Nicaragua's harbors, Moynihan supported him by temporarily resigning his position. His opinion of the CIA did not mellow as the years passed. He poured scorn on the idea that the CIA had been instrumental in America's Cold War "victory." He quoted the English novelist John Le Carré: "The Soviet Empire did not fall apart because the spooks had bugged the men's room in the Kremlin or put broken glass in Mrs. Brezhnev's bath."

The senator from New York challenged not only the competence of the agency but also its enshrinement of the principle of secrecy, a principle that had little utility now that *glasnost*, or transparency, prevailed in Russia. Moynihan's viewpoint is reminiscent of that expressed in a *New York Times* editorial in 1938, which counseled against an expansion of US espionage capabilities on the grounds that transparency prevailed worldwide, making espionage obsolete: "There are few real secrets, military or otherwise, here or elsewhere." In 1995 Moynihan proposed a measure that stemmed, in part, from a similar aversion to secrecy. He called for the abolition of the CIA and for the State Department to take over its intelligence functions. This made him an unwitting Wilsonian, for it would have been a reversion to the days of U-1.[16]

Here, it may be noted that President Clinton subscribed to the doctrine that there should be greater openness in public affairs. Back in 1992, Tennessee's Senator Al Gore (soon to be Clinton's vice president) and CIA director Robert Gates had started the process whereby hitherto secret images taken by Corona and other American spy satellites should be made available for broader scientific research. In 1995, at the height of Moynihan's abolition campaign, Clinton authorized the release of over 800,000 images. They were a boon to the research of a carefully vetted cadre of scientists who over the years contributed to the understanding and prediction of such threats as melting polar ice caps, incipient forest fires, and impending volcanic eruptions. The creation of the CIA's offshoot MEDEA, or Measurements of Earth Data for Environmental Analysis, could be portrayed as a "greening" of US intelligence. At the same time, such activities could be seen as an unwarranted and even opportunistic expansion of intelligence activities into an area—environmental research—where no one was threatening America with secret, hostile intent.

The opening up of green-friendly data was not a political boon for the president. In fact, the sniping season never expired and underwent reinvigoration in 1997, the year of the CIA's fiftieth anniversary. In March of that year, the former CIA analyst Melvin Goodman charged that rigid adherence to the notion that the Soviet Union was an unbending threat had blinded Bill Casey and Robert Gates to the genuine nature of Mikhail Gorbachev's peace overtures and domestic reform. According to Goodman, Gates's successors had since "recycled those high-level officials who contributed to the politicization of intelligence in the first place." Both the Senate and the House Intelligence Committee expressed similar sentiments in their funding reports later in that anniversary year.[17]

It was all to little avail. The CIA survived the onslaughts. Moynihan's campaign against secrecy struck a chord, but his abolition bill never made it to the floor of the Senate for a vote. There were fears that the disappearance of the CIA would permit the Department of Defense to overmilitarize national intelligence. Legislators were coming to realize, in the words of one senior CIA analyst, that Moynihan's was "a simplistic view of intelligence driven by the notion that the Soviet Union was the only reason CIA existed." They heeded James Woolsey's warning: "We have slain a large dragon. But we live in a jungle filled with a bewildering variety of poisonous snakes."[18]

There were countervailing factors that ensured the sound and the fury signified, if not nothing, at least nothing fatal. One of these, as we shall see in the next chapter, was the emergence of anti terrorism in place of the old mission of anti-communism. But there were other helpful trends too.

Mood assessment can help to explain why the CIA was able to retain enough standing in the public eye to enable it soldier on. Reactions to Venona are a case in point. In 1995, Moynihan was given charge of the bipartisan Commission on Protecting and Reducing Government Secrecy. In the course of the inquiry, the erstwhile top-secret details of the Venona program came into prominence. They showed how the program's cryptographers had intercepted and partly decoded Soviet secret messages in the 1940s that yielded information on who had spied for Moscow in the years just before and after the start of the Cold War. The decrypts and translations would be controversial. But at the time of release they were, in effect, a cold shower for American liberals who still protested the innocence of alleged spies such as Alger Hiss and who by extension cast doubt on the integrity of US counterintelligence. The Venona revelations were a fillip for those who upheld institutions of national security like the FBI, NSA, and CIA.[19]

Venona was an arrow in the quiver of America's increasingly influential neoconservatives. Not all neocons embraced the CIA, but beginning with the Reagan presidency the agency had come to be championed more by conservatives than by liberals. The CIA's history came to be seen in a different light. For example, as one biographer put it with the Ames case in mind, the conservative CIA legend James "Angleton's approach to counterintelligence was seemingly vindicated by events in the late 1980s and 1990s."[20]

Consideration of the boom in CIA movies in the 1990s reinforces the point that, in the wide world beyond Congress and liberal academia, the CIA's reputation was far from entirely tarnished. In 1996, the agency attended to its image by appointing Chase Brandon to be its entertainment industry liaison officer. He worked with Hollywood studios and was in demand for contributions that gave movies more "authenticity." However, Brandon did not initiate the upward trend and could not have been responsible for earlier pro-CIA movies like *The Hunt for Red October* (1990) and the distinctly neoconservative *Clear and Present Danger* (1994), both based on Tom Clancy novels. A number of movies about the CIA tended to dwell on tropes like assassination and "rogue" behavior. But behind these entertainments there lurked a continuing regard for the CIA's mission to defend the nation.[21]

Reform was a further factor that helped to keep the critics at bay. The relaxation of tensions at the end of the Cold War had inspired modest changes, even as the attempt to abolish the CIA failed. The reform initiatives of the 1990s took up once again the threads of 1970s reform aspirations.

We can begin with the Bush administration's initiative on transparency. There had long been discontent at the federal government's practice of "classifying" as secret documents that critics claimed should be in the public domain. There were, most people understood, sound reasons for classification. It protected sources and methods. It enhanced objectivity in cases where the release of information might have stirred emotional responses. It shielded the negotiating positions of the US government and prevented harmful speculation based on options that officials had considered without adopting. The process was, however, open to cover-ups and other abuses, and also resulted in unmanageable accumulations of classified documents.[22]

CIA director Gates established a task force on greater CIA openness. Four weeks later, it delivered a report based on consultation with people in government, business, the media, and academia. To the accompaniment of media derision, the report on openness was immediately classified. However, some months later the CIA made it available. The report struck a somewhat defensive note: "Many of those interviewed said the CIA was [already] sufficiently open...we should...preserve the mystique." Yet there was an acknowledgment of "the dramatic changes in the world situation," and the report endorsed the idea that the agency should abandon classification for classification's sake, and be more open about its activities and achievements.[23]

Another reform initiative aimed to improve the contribution and status of minorities. Those who accused the CIA of being the preserve of Ivy Leaguers had overlooked one of its broader limitations—its tendency to favor the employment of white men. Like Jeffrey Sterling, who was first employed in 1993 and would ultimately be convicted of leaking CIA secrets, most of the tiny number of African Americans to be found in the agency felt underrepresented and marginalized.[24]

Aware of another "minority" deficiency, the agency addressed the gender issue. It preemptively publicized the career of Martha Neff, a division chief within the Office of Near Eastern and South Asian Analysis. Interviewed in 1991, Neff praised the opening of a child care facility in Langley in September 1989 and loyally recalled that "George Bush sent me a very nice handwritten note after I briefed him on Lebanon when he was director of the CIA."[25]

The Bush administration's "Glass Ceiling Study" of 1991 recorded that 45 percent of the women employed in the CIA complained of being denied promotions and of being harassed sexually. Such studies encouraged a revolution of rising expectations. Three hundred female workers in the operations directorate now launched a legal suit. Their petition noted that, out of 450 women employed in the CIA, only ten held senior positions. In 1995, the agency settled out of court, tendering almost a million dollars in back pay and offering retrospective promotions.[26]

President Clinton broadened Bush's agenda when in August 1995 he issued an executive order that banned discrimination against homosexuals in matters of security clearance. This was a dramatic change.

As recently as 1980, the agency had not only banned homosexuals but also, in the year when the Democratic Party adopted a gay rights plan, issued a guide on how to spot them. The homosexual had certain behavioral attributes, whose "existence he will deny almost to his last breath." The guide defined, for the benefit of the innocents of the CIA, the meanings of words like "gay" and "straight." It noted, "The question 'Are you gay, straight, or bi?' has been used with marked success in interviews of suspected homosexuals."[27]

According to the self-fulfilling mythology of the spying profession, homosexuals were a security risk because they could be blackmailed into becoming traitors or double agents. Historically, there were precious few examples of this. Opinion is divided on the celebrated case of Alfred Redl, deputy chief of military intelligence on the imperial Austro-Hungarian general staff, 1907–12. Redl was exposed as a spy who had sold extensive details of war plans to Russian military intelligence. On being exposed, he did the honorable thing when presented with a pistol. Ostensibly, his treason sprang from threats to expose his sexuality, but it seems just as likely that he did it for the money.

Once Clinton lifted the ban on homosexuals, the bubble was pricked. There was no point in trying to blackmail gay CIA personnel if their exposure carried no penalty. Established in 1996, the Agency Network of Gay and Lesbian Employees now vigorously campaigned to eliminate gender-orientation discrimination in the CIA.[28]

The struggle for greater equality in the CIA takes its place in the annals of social history. But it is also a strand in the story of secret intelligence. The CIA was at its heart a cognitive institution, and one half of the nation's brains are female, while approximately a tenth are gay. With criticism pouring in about the agency's analytical shortcomings, there was an urgent, practical need for more women and homosexual persons in responsible posts.

As for the racial and ethnic composition of the workforce, there existed an inconvenient truth for white, monoglot English speakers. The world was predominantly composed of people who looked different and spoke differently. The need for diversity in the CIA workforce was evident, and the nation's multicultural composition could have, but had not yet, made its accomplishment a shoo-in. Finally, as we have noted more than once, in a democratic society the CIA needed popular support, and an agency whose appearance approximated to that of the general population was more likely to retain respect and legislative succor.

Moynihan's secrecy commission did not report until March 1997, but in the meantime it prepared the way for a transparency initiative by publicizing the issue of excessive classification. To dramatize its point, the commission held an internal competition for the most absurd secrets. One candidate was a "Top Secret: Eyes Only" document from President Carter's aborted hostage rescue mission. This carefully concealed contribution to national security exhorted pilots to carry no milk in their lunchboxes, as the desert heat would sour it.

A White House official stated that the US government guarded an unrevealed number of secret documents that possibly ran into the billions. A total of 17,000 documents were being classified daily, and the annual cost to the taxpayer of security classification exceeded $4 billion. There was a program of document destruction, but it was erratic, with selections made sometimes for housekeeping reasons and sometimes for reasons that encouraged suspicions of cover-ups.

Against the background of the Moynihan-inspired debate, President Clinton issued executive orders encouraging declassification and discouraging overclassification. The move to greater transparency rested on the logic that it is more effective to guard a smaller number of rationally selected secrets than a whole raft of trivia that nobody respects. Although Gates described "CIA openness" as an "oxymoron," the executive orders applied to the CIA as well as to the rest of the federal bureaucracy. It gradually became easier for historians to examine the agency's past, arguably a development that contributed, on balance, to public confidence.[29]

President Clinton favored the disclosure of the CIA's overall budget, hitherto one of the nation's guarded secrets. This issue occupied the attention of the Commission on the Roles and Capabilities of the US Intelligence Community. The president and Congress in 1995 charged this inquiry, soon to be dubbed the Aspin–Brown Commission, with the task of assessing the post-Cold War status of the intelligence community. The appearance of its report in 1996 met with indifference bordering on disdain. For example, Senate Intelligence Committee chairman Arlen Specter complained that the commission ignored the "biggest problem." The Pennsylvanian Republican believed that this was the undue influence of the agency's unaccountable "old boy network."

However, the commission's budgetary recommendation did have an impact: "The President or his designee [should] disclose the total amount of money appropriated for intelligence activities during the current fiscal year and the total amount being requested for the next fiscal year. The disclosure of further detail should not be permitted." The idea was to reveal a global figure that would dispel suspicions of a "Deep State" with sinister resources, while giving no clues to funding distribution within the CIA and its siblings that might have been helpful to foreign foes.[30]

Gradually, those who wished to know became acquainted with the outline of general intelligence spending, though not specifically what the CIA or other agencies received. The American public learned that in the late 1980s real intelligence spending had increased at an annual rate of 120 percent, a rate substantially higher than that for defense generally. It was a tribute to the standing and persuasive powers of Bill Casey.

The end of the Cold War brought cuts. James Woolsey was obliged to reduce personnel by 24 percent in 1993–4. Agency morale, according to one journalist, at this time sank to a level "lower than Death Valley." The officially released figure for expenditure on intelligence in 1997 was $26.6 billion, perhaps a one-third reduction in real terms from the amount at Casey's prime. But that figure held steady for 1998. In 1999 the CIA's director George Tenet showed a recidivist tendency, withholding the total for that year, but this proved to be a temporary setback, just like the budget reductions of the earlier 1990s.

To sum up, the public could be assured that there was no secret state running away with ever-increasing amounts of their money. For its part, the CIA benefited from the legitimate and open approval of the budget that sustained the agency and its community. The nation had come to accept the principle of post-Cold War central intelligence.[31]

12

Fateful Terror and 9/11

A merica is no stranger to the problem of terrorism. Four American presidents have fallen to assassins, one of them, President William McKinley, the victim of a declared anarchist. In 1954 four Puerto Rican nationalists opened fire on the House of Representatives chamber in the US Capitol, wounding five congressmen. But by the 1990s the nation faced a more systemic and dangerous terrorist threat than it had ever encountered before.

The threat became very evident on 18 April 1983. On that day, a suicide bomber drove his vehicle, carrying a 2,000-pound bomb, into the portico of the US embassy building in Beirut. The ensuing explosion killed sixty-three people, of whom seventeen were Americans, a circumstance that naturally impacted on opinion back home. An Iranian-backed group called Hezbollah ("party of God") phoned to boast about their responsibility.

Of the seventeen American dead, eight were CIA employees. They included the Beirut station chief Kenneth Haas and Robert Ames. The father of six children, Ames was the CIA's chief Middle East analyst. He had built a relationship with the Palestine Liberation Organization and was friendly with the PLO's intelligence chief Ali Hassan Salameh prior to that official's assassination by Israel's Mossad in 1979. Salameh was a marked man who had always walked around with a holstered pistol, and Ames wanted to present him with a new gun as a token of their friendship, until Langley intervened with a veto. The crude Hezbollah bomb attack expunged, in Ames, an American who was working, through the most difficult channels, for a peaceful solution to regional tensions.[1]

The threat epitomized by the Beirut bombing furnished a persuasive reason for the continued existence of the still troubled CIA. Yet

Islamic terrorism would prove to be an even more fateful issue. Fateful for close to 3,000 Americans who died on 11 September 2001 when terrorists hijacked planes to attack New York City's Twin Towers and the Pentagon, and fateful for the CIA itself when a reform act reduced its status in 2004—before ultimately being fateful for the main inspirer of the 9/11 attack, Osama bin Laden.

The Reagan administration had afforded new recognition to the burgeoning menace of terrorism when, in 1986, it established a Counterterrorism Center within the CIA. Under the initial leadership of Duane Clarridge and located on the sixth floor of the Langley headquarters building, the center held forth promise by bridging a problematic chasm, for it drew together analysts and operators in a manner that was unusual for an agency whose security arrangements normally dictated hermetically sealed units. However, the counterterrorism effort lost momentum under the leadership of CIA director James Woolsey (1993–5), who focused on the establishment of a new generation of spy satellites.

There were other legitimate distractions too. The CIA was expected to fight the international narcotics trade that fed gangsterism in supplier nations and ruined so many lives in America. The need to establish expertise in cyberwarfare became very evident toward the end of the 1990s, when there was a Russian-based hacking attack leading to a breach of classified information in the Pentagon and other strategic federal entities. Moonlight Maze, an operation involving national security agency analysts and US specialists working "under the auspices" of the CIA, revealed the extent of the hacking operation in 1999 and thus drew attention to what would be a persistent threat in the following century.[2]

At the Counterterrorism Center itself, there was little appreciation of the fact that the main terrorist threat was now transnational. The center watched the troublesome Saddam Hussein regime in Iraq and the activities of Iranian-backed Shia Muslim groups like Hezbollah, but not the rising tide of jihadism, a holy war conducted independently of the control of any one country.

Events suggested the need for a change of emphasis. On 25 January 1993, just five days after Clinton's inauguration as president, the Pakistani national Mir Aimal Kansi stopped at traffic lights outside the CIA headquarters in Langley, Virginia. He had with him a recently purchased AK-47 assault rifle. He opened fire on stationary vehicles waiting to enter the agency's premises, killing two CIA employees. After the event, he fled to Pakistan. A joint CIA–FBI team hunted him down. He went on trial and was executed in 2002. Though Kansi had no connection with organized terrorism, he came from an area known to be a haven for jihadist fighters in Afghanistan and locally was hailed as a martyr.

Close on the heels of the Langley incident came the first attempt to blow up the Twin Towers. A van filled with explosives was supposed to topple one tower into the other. The detonation killed six people, but the first tower withstood the blast and did not collapse. The perpetrator of that event too was a Pakistani who lacked organized connections. Ramzi Yousef was later hunted down and given a 240-year prison sentence. Yousef was not acting for Pakistan. He was the dutiful nephew of a member of a shadowy organization that would become known as al-Qaeda, an Arabic term meaning "the base" or "the foundation".

In 1995 the Taliban entered Afghanistan's capital, Kabul. The word "Taliban," translated as "students," denoted fundamentalist Islamists who favored strict sharia law. They were the inheritors of the mujahidin ("strugglers") who had resisted Soviet rule with CIA assistance. They aimed to oust the post-Soviet Afghan government that the United States supported and to expel the American military. The Taliban were allied with and gave refuge to al-Qaeda, and their control of large areas of Afghanistan looked ominous.

When John Deutch (1995–6) took over at the CIA in succession to Woolsey, Clinton issued a directive, "US Policy on Counterterrorism." It ordered the agency to hunt down terrorists and to resort to unconventional action where necessary. By now the CIA was aware of the importance of al-Qaeda and its leader, Osama bin Laden. In 1996 Deutch appointed the CIA analyst Michael Scheuer to head a special unit, the Bin Laden Issue Station. Scheuer nicknamed it "Alec," his son's name. Seventy percent of the personnel Scheuer selected were women, and chauvinists renamed Alec "the coven."[3]

The Clinton administration allocated $5.7 billion to the counterterrorism program in 1996. The annual appropriation reached $11.3 billion by 2001.[4] The increase reflected the gathering pace of terrorist incidents. For example, on 7 August 1998 al-Qaeda set off simultaneous explosions outside the US embassies in Dar es Salaam, Tanzania, and Nairobi, Kenya. The truck-delivered bombs wounded thousands and killed 225, including two CIA officers. It was increasingly apparent that al-Qaeda was well coordinated.

The CIA now finished its directorial merry-go-round. George Tenet arrived to serve for a decent spell, from 1997 to 2004. The new director declared "war" on al-Qaeda. President Clinton signed a "finding" (order) giving the CIA authority to assassinate bin Laden. When challenged in a 2006 interview with the conservative Republican contention that he had not done enough, he retorted that he had done plenty, unlike the succeeding George W. Bush administration: "What did I do? What did I do? I worked hard to try to kill him. I authorized a finding for the CIA to kill him. I got closer to killing him than anybody has gotten since. And if I were still president, we'd have more than 20,000 troops there trying to kill him."[5]

The 1970s ban on assassination had been a liberal measure that had been partially eroded by the CIA-sponsored Contra manual issued in the conservative Reagan presidency. It had now been overturned by the latest liberal incumbent of the White House. On 20 August 1998 the CIA made its first attempt on bin Laden's life. US Navy ships in the Arabian Sea launched a Tomahawk cruise missile attack that demolished a base near Khost, Afghanistan. Not for the last time, the agency had got its bin Laden intelligence wrong, for the Saudi was absent from the dead.[6]

If, by the end of the century, the American government and the CIA were alert to the menace of Islamic entities like al-Qaeda and were already focused on the threat posed by Osama bin Laden, why was the 9/11 attack such a "surprise"? In 2004 a bestselling book appeared that pointed to a long-term, as opposed to immediate explanation. At first, the author of *Imperial Hubris* styled himself as Anonymous, but the public soon learned that he was Michael Scheuer, the CIA analyst who had headed the CIA's Bin Laden unit from 1996 to 1999.

Scheuer argued that 9/11 was a "tragedy" in being the outcome of an almost willful misapprehension of the nature of America's declared foe. To portray bin Laden as an arch villain was myopic. In a later work, Scheuer maintained that the Saudi militant had some of the characteristics of a great man. He was pious, poetic, charismatic, and brilliant. In Scheuer's view, bin Laden devoted himself to causes with which the majority of the world's 1.3 billion Muslims could associate: opposition to the Arabian tyrannies, to Israeli expansionism, and to US imperialism.[7]

One of the subheadings in *Imperial Hubris* was "Semantic Suicide: Fighting Terrorists when Faced by Insurgents." Scheuer argued that it was wrong to assume that Islamists disliked American values such as democracy—"they hate us for actions not values." America had given offense because it had preached the values of democracy, while acting to protect its oil interests by buttressing tyranny in Saudi Arabia and other Middle Eastern countries.[8]

There can be no excuse for the murderous actions of bin Laden and terrorists like him, but Scheuer did have a point. Terrorists' dispatch and inspiration of suicide bombers, often vulnerable individuals suffering with psychological issues, cannot be justified by the tenets of any respected faith or moral code. But the bin Ladens of the modern era could not have thrived without the sympathy and support of local populations. If someone commits a terrible deed in the face of terrible oppression, it is only natural to see that deed as "understandable," if not excusable. As the Martinique anti-colonial theorist Frantz Fanon noted, yesterday's bandit is today's freedom fighter, and tomorrow's revolutionary hero. The United States itself was a revolutionary nation, but the lessons of that revolution were all too easily forgotten in confronting a form of terrorism which did not relate to nationalism.[9]

Scheuer portrayed American policy toward Israel as particularly counterproductive. When he lost his CIA post in 2004, it was for expressing this view. Official policy dictated that Israel was an essential democratic partner in an unstable oil-rich region. Though socialist at the time of its inception, Israel was now capitalist and neoconservative, an additional attraction for an increasingly powerful group of US policy influencers. But Scheuer thought it was a strategic error to be rigidly devoted to Israel.

Sentiment on Israel had changed since the CIA's early years, when the agency's Arabists had sympathized with Middle Eastern anti-imperialists. Tenet, for example, was now patronizing in tone when writing about the Palestinians at whose expense Israel was expanding. To take an example from his autobiography, the CIA director contended that an election victory by the Hamas party, which insisted on Palestinian rights, "was disastrous for the peace process" and that Middle Easterners should have education prior to democracy, to which they should be "allow[ed]" to proceed "at their own pace."[10]

Scheuer let loose at an American policymaking establishment that he thought pandered to outmoded imperialist ideas and had other problems too:

We dithered fatally because our leaders and their analysts…refused to accept the shift in the terrorism problem from one of state sponsors and their sponsors to one of Islamist insurgents.…Also constraining our counterterrorism measures…is the tragic reality that American lives mean little when weighed against U.S. officialdom's concerns for the opinions and reactions of foreign, especially European, governments [and the danger of] scorn from the [*Washington*] *Post* [that might] delay the next step up the career ladder.[11]

Scheuer's logic did not sweep all before it. His revelation that politicians were political was less than earth-shaking. One UK journalist accused the former counterterror analyst of pumping up the abilities of bin Laden as an excuse for his own failure to neutralize the terrorist. Yet the idea that Europe "lured" the United States into adopting colonialist attitudes has had some currency, and the notion that America acquired a "diabolical reputation throughout much of the Arab world" is widely accepted. Scheuer's hypothesis on these underlying causes of 9/11 does carry some weight.[12]

One veteran intelligence investigator has drawn attention to a "lingering question" asked in the wake of 9/11, namely, "whether a stronger DCI, as recommended by the Aspin-Brown Commission, could have steered the nation's intelligence services toward a more effective gathering of information about terrorists—especially al Qaeda."[13] In its report, the commission had recommended a formula for the exercise of power through the purse strings: members of the intelligence community should be obliged to report their budgets to the CIA's director wearing his director of central intelligence hat, and he should be empowered to arrange "tradeoffs" leading to "resource management across the Intelligence Community."[14] The exhortation proved to be in vain, and non-implementation of investigative wisdom deserves to be considered as a further long-term cause of the non-anticipation and non-preemption of 9/11.

Lack of intelligence coordination had been a problem at Pearl Harbor, the event that had focused the collective mind of Congress on establishing the CIA in 1947. The arrangement whereby the CIA boss would have an additional, coordinating role had been established in

1947, yet was a bone of contention from the start, with rival agencies such as the FBI reluctant to surrender their prerogatives. CIA director John Deutch recalled that, when he tried to coordinate, the secretaries of defense, state, and justice all "hated it."[15]

When George Tenet took over at Langley, he strove like his predecessors to harmonize the efforts of his security colleagues. For example, one of his directives, effective 1 July 1999, sought to coordinate the intelligence community's information technology effort. This was important in combating terrorism, as al-Qaeda and other jihadists were increasingly resorting to digital communication. The initiative was also far-sighted in another way, as cyberwarfare loomed on the horizon. However, the wording of the directive betrayed caution, nervousness, and even impotence in dealing with colleagues, especially those in the Defense Department. The directive timidly stated that it did "not affect the authorities, responsibilities, and restrictions relating to components of the IC and the Department of defense (DoD), that are set out in existing statutes, executive orders, and policy directives."[16]

President Clinton was prepared to point the finger of death at bin Laden and authorized an increase in the counterterrorism budget. But his administration did not invest political capital in the reform of America's intelligence apparatus. It could not, as a cover-up scandal eviscerated its potency. Clinton discovered to his cost that engaging in fellatio with his young intern Monica Lewinsky and then lying about it was not a good idea. The House of Representatives impeached him in the fall of 1998 and, although the Senate acquitted him, the scandal was a time-consuming obsession and a distraction.

While floundering in disgrace, Clinton tried to handle the various crises and challenges that arose during the remainder of his presidency. At home, the Columbine High School shootings that killed thirteen and the ensuing debate about the power of the National Rifle Association sapped his leadership energies—as did, abroad, the US peacekeeping intervention in Kosovo and the beginnings of an abortive effort to bring peace to the Israel–Palestine issue.

As a security issue, a new spy case captured headlines in a way in which administrative reform never could: in December 1999, the Los Alamos scientist Wen Ho Lee was charged with betraying atomic secrets to China. When the charges did not stick, there was a debate about media racism that diverted attention from counterterrorism.

In October 2000 al-Qaeda posted a reminder from Yemen when its operatives bombed the guided missile destroyer USS *Cole* in Aden harbor, killing seventeen Navy sailors. But by this time the United States was in the final throes of a presidential election. To sum up, a weakened president lacked the focus and the authority to address the centralization reform recommended by the Aspin-Brown Commission.

House of Representatives Speaker Newt Gingrich was Clinton's tormentor-in-chief during the impeachment proceedings and helped to emasculate the president's power to achieve reforms. The Republican from Georgia was committed to the goal of strengthening national security in his own way. Two years after the setting up of the Aspin–Brown Commission, he prevailed on Clinton to charter, under the more conservative auspices of the Department of Defense, a US Commission on National Security for the 21st Century. Known as the Hart–Rudman Commission, the new review worked until January 2001 on a wide-ranging appraisal of the nation's defense needs and priorities.

Hart–Rudman identified the threat that terrorists posed to what it termed the "homeland" as the nation's number-one security problem. It recommended "the creation of a new independent National Homeland Security Agency (NHSA) with responsibility for planning, coordinating, and integrating various government departments involved in homeland security." In 2015 its co-chair, Gary Hart, looked back at its efforts. "No one listened," he said. While it is open to question whether a new department would have been able to redress the issue of central authority, it seems reasonable to reiterate the point that the failure of reform may have contributed to the failure of security on 11 September 2001.[17]

In the years, months, weeks, and even days leading up to 9/11, the CIA issued numerous warnings. The "Alec" unit noticed that bin Laden was intent on using airplanes to inflict a severe blow. It picked up speculation that the Eiffel Tower in Paris or CIA's headquarters in Virginia might be the targets. In the event, nineteen al-Qaeda suicide attackers would indeed hijack four airplanes. Two of them rammed into the Twin Towers, a third hit the Pentagon, and the fourth, in spite of being counter-hijacked by its brave passengers, crashed in Pennsylvania with no survivors.[18]

George W. Bush had assumed office as the nation's forty-third president on 20 January 2001. CIA director Tenet had intimate access to the new chief executive. He personally delivered morning briefings to the president on forty occasions between Bush's inauguration and the calamity of September. He also briefed other senior members of the Bush administration. On the afternoon of 10 July, Tenet received news that made his "hair stand on end." He reached for the white telephone in his office that gave direct access to National Security Adviser Condoleezza Rice and asked for an urgent audience. At the ensuing meeting, Rice learned that there would be "a significant terrorist attack in the coming weeks or months." Tenet rejoiced in the knowledge that he "had gotten the full attention of the administration."[19]

On 6 August Tenet as usual carried with him to his White House meeting a document known as the President's Daily Brief. On this day, its title was "Bin Laden Determined to Strike in US." It noted that in television interviews a few years ago, bin Laden had hinted that his followers would heed the example of the earlier Twin Towers bomber Ramzi Yousef. It further revealed that the CIA and FBI were investigating a call to the US embassy in the United Arab Emirates, "saying that a group of Bin Laden supporters was in the US planning attacks with explosives."[20]

The problem was that the CIA could not specify when and where the impending attack might occur or who the prospective attackers were. The agency had failed to penetrate al-Qaeda, and its informants in Afghanistan were no help. It had the means to intercept messages, but lacked specialist translators. One CIA veteran, an accomplished linguist, asked how many agency personnel spoke Pashto, the language predominantly spoken by the Taliban who were sheltering bin Laden. Answering his own question, he said, "None."[21]

There were a number of communications problems. At ground level, the approximately 200 field agents dispatched to hunt down bin Laden were said to be reluctant to "take direction from the ladies" in the "coven."[22] There was a similar charge of inaction at a higher level. The veteran journalist Bob Woodward accused Tenet of having failed to exploit his ample opportunities to pressure President Bush into taking some kind of preemptive action.[23]

If Tenet had really wanted a response, he could have resigned in protest at the president's inaction, causing a fuss that would have forced Bush to concentrate on the problem of an imminent attack. But the CIA is supposed to be apolitical, and Tenet liked his job.

In defense of the CIA, one could point to other deficiencies in the nation's homeland security arrangements on the eve of 9/11. After the dreadful event, these deficiencies would receive exhaustive attention from Congress, the press, and numerous investigators. But spreading the blame would not save the CIA from a fateful slump in its standing and role in national intelligence.

13

The Great Diminishing Reform
Act of 2004

The Intelligence Reform and Terrorism Prevention Act of 2004 does not stand out as one of the great events of the administration of President George W. Bush (2001–9). For the CIA, however, it was the biggest structural and mission change since 1947.

Any discussion of the causes of the reform leads straight into the realm of controversy. All agree that it was a response to the 9/11 attack. But critics charged that it was politically driven. They argued that the Bush administration had failed to protect American citizens on that fateful day in the fall of 2001 and sought a scapegoat for its own failure. That search led to the targeting of the CIA, which was charged with dereliction of its duty of preventive intelligence and punished with demotion for offenses it did not commit.

CIA partisans and their allies insisted on the unfairness of the charges against the agency. Some pointed to the White House's poor handling of intelligence. At the time of the 9/11 attack, Condoleezza Rice was President Bush's national security adviser. When a professor at Stanford University in the 1980s, this only child of a college dean had lectured students on the reason for the Pearl Harbor intelligence failure. But had she learned her own lessons? As she aptly recalled, "It's one thing to read about it and quite another to be in control, maybe the central character in the drama."[1]

On Rice's watch there had been a lack of receptivity to clues. A month after she failed to attend a 5 July 2001 meeting of counterterror officials that she herself had convened, CIA Director George Tenet issued his warning of an acute danger of an attack by al-Qaeda. Rice later remarked that this 6 August report cobbled together old and

inconclusive information. She contended that the very fact that the
president had had to ask for a report on the terrorist threat showed that
the CIA was not doing its job. Not everyone believed this sophisticated defense.[2]

While some defenders of the CIA pointed diversionary fingers at
Rice, others argued that the administration had failed to take a wider
view of what caused the terrorist problem. Michael Scheuer, it will be
recalled, had a challenging interpretation of events: "Washington's
maintenance of a policy of status quo toward the Moslem world and
its more or less constant green light for Israel's actions against the
Palestinians would have resulted in more young men volunteering for
jihad even if bin Laden did not exist."[3]

In our review of the causes of the 2004 legislation, we shall consider
further issues that are not so much political controversies as historical
discussion points. One of these concerns the timing of the decision to
demote the CIA in America's intelligence hierarchy. The drastic reform
came at a time of national peril, a departure from precedent. The other
issue, more speculative in nature, is the part played by the father–son
relationship between the first Bush president, George H. W., and his
heir in the White House, his son George W. We shall also review the
consequences of the Reform Act. For in the short term at least, the
standing of the CIA declined. Leon Panetta recognized that decline.
When he became director of the CIA in 2009, he said that his "first
goal" was "to restore its standing with the American public and political leadership."[4]

In instigating reform, the Bush administration was at first constrained by the fact that public opinion did not immediately turn
against the CIA in the wake of 9/11. Just after the event, people stood
outside the White House patriotically shouting "CIA! CIA!" There
was a brisk trade in CIA T-shirts and many thousands of applications
to join the agency. On Capitol Hill, legislators competed to display
their eagerness to increase the intelligence budget. Official inquiries
into 9/11 and the issue of intelligence failure did not at first support
the idea that the CIA should be demoted.

The time was not right to accuse the CIA. Instead, President
George W. Bush acted as if he expected the agency to solve the terrorism crisis. He deployed what was, after all, the only agency of the
federal government with a capability to attack al-Qaeda.[5] Attack being
an effective means of defense under the circumstances, the White

House encouraged the agency to engage in a manhunt, backed up by paramilitary action, in parts of the world that harbored al-Qaeda operatives. To that end, in October 2001 US forces invaded Afghanistan, a country accused of continuing to harbor Osama bin Laden. The invasion enabled the quashing of the terrorist group in that country, permitted the CIA to establish local intelligence-gathering facilities, and also developed into a mission with further objectives: billions of US dollars were invested over the next two decades in building new schools, promoting democratic infrastructure, and helping Afghan women in their quest for greater rights.

In the more immediate term, the administration soon articulated a need to improve the cohesiveness of an intelligence community that had not "connected the dots."[6] The CIA, FBI, and other sources had provided clues that might have led to the prediction and preemption of the 9/11 attack if pieced together. Turf wars and a commitment to "stovepiping"—the security principle whereby different spy components keep each other in the dark as a precaution against foreign penetration—had prevented the collation from taking place. In June 2002 Bush picked up on a recommendation of the Hart–Rudman Inquiry, the brainchild of Newt Gingrich in the previous century. The president proposed a new cabinet-level Department of Homeland Security. He said it would "complement" the intelligence-gathering functions of the CIA and FBI, and analyze the data each of those agencies supplied.[7]

Criticism of the nation's intelligence performance from both ends of the political spectrum encouraged the Bush administration to reform. Conservatives objected to liberals' imposition of a ban on "racial profiling." The journalist Mark Steyn excoriated the FBI's failure to arrest one of the potential suicide mission pilots: "In August 2001, invited to connect the dots on the [Zacarias] Moussaoui file, Washington bureaucrats saw only scolding editorials about 'flying whilst Arab'." Steyn's headline, "Stop Frisking Crippled Nuns," summed up conservative fatigue over the constant fealty to political correctness that they thought had hampered the counterterrorist effort.[8]

Liberals countercharged that American intelligence was too intolerant of foreign cultures and languages. Critics revived complaints about the linguistic shortcomings of the CIA, especially in the case of Arabic. *The Washington Post* seized on that failure in regard to two Arabic-language messages in particular. Sent on 10 September 2001, they had warned of the attack the next day, stating "Tomorrow is zero hour."[9] The United States was home to more than a million Arab Americans at this time, most of them Christian and not likely to be a security risk. Yet nobody could be found within the CIA or its sibling agencies who could instantly translate those messages.

Official investigations sharpened the knives of those who wanted to blame the CIA. The joint congressional inquiry into the terrorist attack reported in December 2002. It faulted the CIA for failing aggressively to "watchlist" known members of al-Qaeda. Republican Senator Mike DeWine of Ohio told the inquiry there should be a "new OSS" to hunt down terrorists. The joint inquiry listened to witnesses who advanced the pros and cons of replacing the post of director of national intelligence with a new post that would not be tied to the CIA. However, the inquiry did not concentrate on the agency and set forth other shortcomings of the intelligence community, for example, pointing to the "Wall" created by the founding legislation of 1947, which meant that the domestically focused FBI could cite legal reasons for not sharing information with overseas oriented agencies like the NSA.[10]

In November 2002 the president and Congress established a wider-ranging inquiry, the National Commission on Terrorist Attacks upon the United States. This inquiry endorsed the unseating of the director of the CIA from his role as director of national intelligence, thus rubber-stamping George W. Bush's effort to single out the CIA as the scapegoat for 9/11.[11]

By the time of the terrorism commission's final report in July 2004, another event had taken place that would further undermine the standing of the nation's premier intelligence agency and thus prepare the way for the Reform Act of December. On 20 March 2003, US forces invaded Iraq. Although the operation did not this time have UN approval, there were many who at the time thought it was a great idea to get rid of that mini-Hitler of the Middle East, Iraq's president Saddam Hussein. The autocrat's regime had used chemical weapons against Kurdish nationalists and killed or "disappeared" an estimated 250,000 of its own citizens.

When Baghdad succumbed to the military onslaught, Hussein went into hiding. US forces mounted a search, and eventually ran him down and arrested him. An Iraqi special tribunal found him guilty of crimes against humanity, and the toppled dictator was hanged on 30 December 2006. In spite of this apparent success, President Bush's justification of his invasion spelled disaster for the CIA. The president believed it was not enough to say he was going to remove a despot. After all, the world contained a few other despots, and people might ask why he was not going after them all. Bush was convinced it was necessary to persuade the American people that Hussein's Iraq was a threat to US national security. Saddam Hussein, he announced, had weapons of mass destruction that could be used against the United States. Thus was born the "WMD" controversy.

Since becoming president of Iraq in 1979, Hussein had indeed armed his country with chemical and biological weapons and had pursued a program designed to equip Baghdad with a third type of WMD, the nuclear warhead. Under pressure from the international community he had scrapped all these programs. Later, because of suspicions that he was reverting to old habits, the United Nations sent in inspectors, who, in the period from November 2002 to March 2003, found no evidence of infractions. Hawks in the Bush administration argued that the absence of evidence was no proof of innocence. They pointed to Hussein's instincts for concealment—he ordered the destruction of paper trails and other evidence sought by UN inspectors. In a retrospective analysis of its own shortcomings, the CIA accepted Hussein's innocence in the matter of WMDs, but noted that, because of his fetish for secrecy, Iraq's ruler had destroyed the very evidence that would have proved his innocence.[12]

Administration hawks Deputy Director of Defense Paul Wolfowitz, Vice President Dick Cheney, and Secretary of Defense Donald Rumsfeld pressed their case that Iraq had WMDs. Tremendous pressure was brought to bear on the CIA's WMD unit, whose members knew full well that Hussein did not possess the alleged weaponry. The CIA's former counterterrorism chief Vincent Cannistraro charged that Vice President Cheney and his aide Lewis ("Scooter") Libby descended on Langley and bullied mid-level analysts to find evidence that did not exist. One officer who wished to remain anonymous complained, "No one is willing to say the emperor has no clothes.... Your job is just to salute and say OK."[13]

Those who wanted war were circulating unreliable information. For example, informant "Curveball" (Rafid Ahmed Alwan al-Janabi) supplied allegations on the basis of which the CIA and White House proclaimed in October 2002 that Hussein had chemical and biological weapons and would be able to go nuclear by the end of the decade, or sooner with foreign assistance. The litany of misinformation persuaded—or enabled—basketball fan George Tenet allegedly to inform President Bush that the evidence on WMDs was a "slam dunk." When that story later leaked to the press, Tenet said the White House circulated the phrase because it was trying to "shift the blame from the White House to CIA" for committing the WMD error. But he did not deny using the phrase.

In addition to citing CIA confirmation of WMD, the president was able to refer to support from Prime Minister Tony Blair's UK government. In February 2002 the administration had given credence to the myth that Iraq was seeking a supply of yellowcake uranium from the African state of Niger. It would have helped Iraq to develop an atomic bomb. A year later, Bush's January 2003 State of the Union address contained the infamous "sixteen words", a sentence that would later embarrass the governments of two nations: "The British government has learned that Saddam Hussein recently sought significant quantities of uranium from Africa." In the following month, Tenet sat at General Colin Powell's shoulder when the secretary of state assured the UN Security Council that Hussein had the weapons.[14]

Condoleezza Rice described the intelligence consensus as follows:

...it was the unanimous view of the U.S. intelligence community that [Saddam Hussein] had reconstituted his chemical and biological weapons programs. All but one believed that he was reconstituting his nuclear weapons capability as well....In 2001 we had failed to connect the dots. We would not do so again.[15]

The decision to invade Iraq soon came to be seen as a major mistake. In part, this was because of an associated ill-judged choice. The Bush administration set out to dismantle the political and administrative structure of the subjugated nation on the grounds that the ruling Ba'ath Party was overcommitted to socialism and the public ownership of oil resources, and that the Party was too much associated with minority Sunni Muslim rule in Iraq, a predominantly Shia Muslim society. The stripping out of experienced administrative personnel created a power vacuum that a new generation of jihadists exploited. The decision further demoralized CIA personnel. Sam Faddis, a CIA officer with experience on the ground in Iraq, put it this way: Washington fired "every single man in Iraq on whom the occupying forces need to rely for the control of that deeply fragmented society. In the process, America not only lost their services but made them enemies."[16] American troops had to commit to years of conflict in Iraq, and the United States cemented its position as the bête noire of the Muslim world.

The CIA's standing in Washington and beyond crumbled because of the ill-conceived decision to invade. The deteriorating position on the ground in Iraq further increased the shock, as did new revelations. Doubts began to form when US occupying forces searched for WMDs and found nothing. Then on 6 July 2003 the *New York Times* published an op-ed by career diplomat Joseph P. Wilson. In 2002 the CIA had dispatched Wilson to Niger to investigate the charge that Hussein was acquiring uranium. He now revealed, in his op-ed, that Hussein had been doing no such thing.[17] To punish the truth-telling Wilson, members of the Bush administration victimized his wife, Valerie Plame. Plame was an officer of the CIA who specialized in nuclear arms control. Bush's colleagues leaked her identity to the journalist Robert Novak, who named Plame in a published article, thus blowing her cover and ending her career in the CIA.

The Plame affair dragged on. Vice President Dick Cheney's chief of staff "Scooter" Libby fell under suspicion as the person who had leaked Plame's identity. He appeared before a federal grand jury in March 2004, and in 2007 would be convicted of perjury and obstruction of justice. Bush commuted his sentence, and President Donald Trump would later issue a pardon. Libby's implied exoneration mirrored the partisan politicization of intelligence that had been at the back of the Iraq War.

CIA Director Tenet was a more prominent person to take the rap. Tenet told a meeting at Georgetown University on 5 February 2004 that the CIA had set up an internal inquiry into the intelligence background of the decision for war, a broad hint that all had not been well. His attitude was defensive and resentful, yet recondite. He sniped at members of the Bush administration for floating inflammatory ideas such as the notion that Hussein had plotted 9/11. He chided Condoleezza Rice for not heeding CIA warnings about 9/11 and for failing to oppose "de-Ba'athification". Yet he took his share of responsibility. "Yes," he conceded, "we at CIA had been wrong in believing that Saddam had weapons of mass destruction." However, Tenet insisted that members of the Bush administration had scapegoated him: "I was the guy being burned at the stake."[18] On 3 June 2004 George Tenet resigned after seven years at the head of the CIA.

Tenet knew about the doubts in the intelligence community, but chose to back the fake message promulgated by President Bush and his hawkish advisers. The CIA director was proud of being the son of "working-class parents" in Queens, New York. Perhaps he was a little too anxious to cling to the high status he had achieved in life, culminating in the directorship of what had been a prestigious agency.[19] To have questioned decision-making in the White House would have brought a premature end to his cherished career.

Though each blamed the other, George Tenet was in the same situation as Condoleezza Rice. Rice did not really have Tenet's humble background. But, a descendant of slaves, she may have been over-acquiescent in warmongering because she was in perpetual awe of her own achievement, an attitude summed up in the title of her memoir, *No Higher Honor*. Her loyalty paid off. Bush promoted Rice to secretary of state in January 2005, adding to the impression that his administration had not been at fault and that she had performed well.

Social mobility so often leads to conformity. But not every CIA director has been a climber, and the truth is that no CIA director has ever mounted a full-on political challenge to the White House. When CIA directors resigned, they did so in disgrace, not protest.

Not a few CIA personnel had their doubts about Tenet. But the departure of their leader with his tail between his legs was still a blow to their morale and to the agency's reputation. On top of this, two reports in July 2004 left the agency's self-esteem in tatters. The first was the Senate Intelligence Committee's *Report on the U.S. Intelligence Community's Prewar Intelligence Assessments on Iraq*. The report slammed the October 2002 intelligence estimate on Iraq WMDs as "overstated." The committee concluded that the intelligence community, which the director of the CIA was supposed to coordinate, suffered from "a broken corporate culture and poor management, and [the problem] will not be solved by additional funding and personnel."[20]

Less than two weeks later, the 9/11 Commission reported. It concentrated its fire on the spooks, not the politicos. It stated that the director of the CIA was no longer able to perform effectively as the overall director of central intelligence. He had too many hats to wear and was not equal to the task of managing complex components of the intelligence community like the National Security Agency. *The New*

York Times observed that the commission shredded the performance of
the CIA over its handling of pre-9/11 warnings, while the FBI and its
director Robert Mueller emerged with much more credit.[21] The commission concluded that there was a "need to restructure the intelligence community."The CIA should be stripped of its overall leadership
role, should surrender paramilitary operations to the military, and
should concentrate on intelligence. The commission called for the
appointment of a new national intelligence director with a "small" staff
of a few hundred.[22]

Its denigrators complained that the 9/11 Commission was biased.
Philip Zelikow, the commission's executive director, had worked with
the Bush transition team and had served on the administration's
Foreign Intelligence Advisory Board. Critics claimed that he was too
close to the administration and took advantage of the opportunity "to
mold the [commission's] findings in a way that exonerated the administration."[23] But, whatever the truth of the matter, the report of the
9/11 Commission bolstered what was becoming a common perception, that the CIA was to blame. It strengthened the case for the passage of the 2004 Intelligence Reform and Terrorism Prevention Act.

The Reform Act was unusual in two respects. First, it occurred in a
period of international tension, when there was acute fear of the terrorist threat. Historians of the intelligence reform cycle have observed
that, in periods of crisis, people typically rallied behind the CIA, an
institution that was symbolic of patriotism and the national will to
overcome all enemies.[24] Such was the case in the immediate wake of
9/11. So what changed? For a full explanation of the reform initiative
there is perhaps a need to look beyond the lessons of 9/11 and WMD.

It is possible that the reform related, in part, to the dynamics of
father–son relationships. We know from their correspondence that
John Adams, the second president of the United States, went to great
pains to shape the outlook of his son John Quincy Adams, the sixth.
According to George W. Bush, the forty-third president, his father
George H. W. Bush, the forty-first, had an influence too. In his memoir he paid tribute to his father: "I never had to search for a role model.
I was the son of George Bush."As to the CIA, he wrote: "I had great
respect for the Agency as a result of Dad's time there [as director,
1976–7]." When deciding who should lead the agency in the wake
of his election victory in 2000, the young Bush consulted his father:
"I asked Dad to sound out some of his CIA contacts."The feedback

persuaded him to reappoint President Clinton's CIA director, George Tenet.[25]

It is conceivable that some of Bush Jr.'s actions reflected a desire to complete his father's business and exonerate Bush Sr.'s decisions. When the American-led coalition liberated Kuwait in 1991, it had driven Iraq's armed forces into full retreat. George H. W. had been criticized for calling a halt to the GIs' advance, missing the opportunity to capture Baghdad and remove Saddam Hussein from office. When George W. Bush set out to finish the job, he received his father's approval. "You are doing the right thing," father said to son.[26]

But the father–son relationship was evolving. Perhaps because Bush Jr. had early problems with alcohol, his younger brother Jeb, governor of Florida from 1999, had at first seemed to be the political heir apparent. There may or may not have been a trace of resentment in George H. Bush's outlook, given those circumstances. Certainly, there was a desire to cut loose from parental control: "My goal was to establish my own identity and make my own way."[27] He had trusted his father's judgment when asking Tenet to continue at the CIA at the outset of his presidency. After 9/11, he could inferentially *blame* that appointment on his father. His sidelining of an agency that his father had once directed was a way, perhaps, of breaking free from the paternal legacy.

The Intelligence Reform and Terrorism Prevention Act established a new overall intelligence chief. It was not a novel idea.[28] There had been several earlier proposals to establish a "tsar" who would be independent from established agencies, including the CIA, and who would have the authority to guide and command the activities of all. Turf wars and political expediency had defeated them—until 2004.

Those turf wars still raged. The CIA naturally hated a proposal that would have stripped its director of his dual mandate. But special circumstances now prevailed. The Bush administration's narrative about 9/11 and WMD demanded a reduction in the CIA's role. The approach of the 2004 presidential election gave special urgency to that narrative. Bush needed to be able to communicate to the nation's voters that the CIA was at fault and that he had fixed the intelligence problem.

There were complaints that the reform was being rushed through to accommodate George W. Bush's electoral needs. In the confusion of the rush, the Department of Defense with its powerful allies in Congress managed to have a clause inserted that the new director of national intelligence should not have the power to "abrogate the statutory responsibilities of the heads of departments." Former Defense Intelligence Agency chief James R. Clapper remarked that the clause "effectively neutered the legislation."[29]

In principle, the Reform Act gave the new director of national intelligence the power to manage the nation's entire intelligence budget and to hire and fire the heads of intelligence agencies, the CIA included. The CIA would not in future have any control over the central direction of intelligence, and the newly created intelligence tsar could not, when in office, hold the directorship of the CIA. The law established a new National Counterterrorism Center that would take over some of the CIA's functions without being answerable to it. It was a major demotion for what had hitherto been the nation's premier intelligence agency.

Senior officials in the CIA were bemused by the deliberations leading to the legislation. They just did not understand what was happening to them, or why.[30] Art Hulnick, a veteran of and expert on the agency, questioned whether wholesale reform should occur in response to the "threat du jour."[31] The respected former judge and political commentator Richard A. Posner saw the legislation as premature. The framers of the Reform Act should have waited for the outcome of the deliberation of the Commission on the Intelligence Capabilities of the United States Regarding Weapons of Mass Destruction, established in February 2004. (In 2005, this commission would criticize intelligence processes, but the intelligence politicization issue was not part of its remit). The text of the new law was clear to Posner on only two points: it blunted the "centralizing thrust" demanded by the 9/11 Commission and the CIA was to be weakened. In retribution for its real or invented deficiencies the CIA would "take the hit."[32]

Amidst the welter of criticism, it is appropriate to pause for thought. Certainly, the Reform Act reflected opportunism and false reasoning. But perhaps it was the right reform for the wrong reasons. The virtues of having a central evaluating process that was separate from and senior to the collectors of evidence had been appreciated ever since the presidency of Woodrow Wilson, and that state of affairs had now been achieved. However, making the system work was, as ever, reliant on the personalities involved.

James Clapper, Director of National Intelligence (DNI) between 2010 and 2017, later noted the early weakness of the organization he had headed. It was perhaps because of that perceived weakness, he suggested, that former CIA chief Bob Gates had turned down the DNI job when it was first on offer. In April 2005, the career diplomat John Negroponte instead became the first DNI.

The DNI's potency would vary over the years according to circumstances and personnel. For there was a point of systemic weakness. As Loch K. Johnson, a veteran of intelligence commissions and oversight committees, observed, satisfactory intelligence arrangements could now be made, but "they lacked the permanency of law."[33]

Nevertheless, faith in the notion of intelligence tsars endured and was legitimized by imitation. Established in 2019, the Cyberspace Solarium Commission recommended, in the words of *The Economist*, "a national cyber-director within the White House, a co-ordinating role much like that of director of national intelligence, which emerged from the 9/11 commission's report."[34]

Stripped of its higher responsibilities, the CIA meantime developed in ways that created additional questions about its standing. As we shall see in Chapter 15, following the passage of the Reform Act, the agency would become more than ever associated with international lawlessness. Analytical responsibility passed to Nick, or the National Intelligence Council.

14

Estimating Anew and a Military Turn

The 2004 Reform Act marked the end of President George W. Bush's first term in office but also of a longer phase in American history that went back to 1947. Before 2004, policymakers had acknowledged the ascendency of the CIA in both the analytical and the covert operational realms. After 2004, the analytical supremacy was gone, and military poaching assisted by intense criticism threatened to weaken the CIA's grip on covert action.

There are two common perceptions of what happened to the CIA in Bush's second term. The first focuses on covert action and portrays the CIA as sinking in a mire of ill-fated operations, before being flayed for undertaking those ventures in such a brutal manner.

But we shall in this chapter deal with the second common perception. This is that the 2004 law forced the CIA to yield the analytical high ground in an unsatisfactory manner. In the words of one historian, "The new office of the [Director of National Intelligence] was a political fix that muddied lines of authority, touched off turf battles, and confused everyone."[1]

A powerful torrent of CIA advocacy has hammered home that viewpoint. There is, however, a countervailing interpretation of what happened. For an exposition of that interpretation we can turn to one of the quieter figures of the Bush administration. "I try hard not to be colorful," Thomas M. Fingar remarked. Fingar was an experienced intelligence analyst whose rise to prominence occurred beyond the CIA umbrella. In the 1970s he had been an Army linguist and analyst. He subsequently enjoyed stints at the State Department's Bureau of Intelligence and Research, the INR. At the INR he developed pride in its independence, especially from the CIA. He recalled that, within the intelligence community, "deferring to CIA judgments" was a "long-standing phenomenon before 2005," but "typically not all analysts" agreed with those judgments.[2]

When John Negroponte became the first director of national intelligence in 2005, he appointed Fingar to be deputy director of national intelligence for analysis and, concurrently, chairman of the National Intelligence Council. At first, Fingar had opposed the creation of a director of national intelligence. He felt it might encroach on the prerogatives of the INR, of which he was then assistant secretary. He wrote also of reservations that the proposed integration would lead to "groupthink and domination by a single big agency with the initials CIA." Once on board with the new bureaucracy, he changed his mind: "I have become convinced that the existence and exercise of DNI authorities are absolutely essential to transforming the Intelligence Community."[3]

Fingar's job was to "lead the analytic transformation effort" mandated by the Reform Act.[4] Negroponte asked him whether he would be prepared to take charge of overhauling the President's Daily Briefs. After brief hesitation, Fingar recalled, "Ego whispered in my ear that it would be cool to oversee the PDB so I simply responded, 'OK'." The Reform Act had not specified a recipe for future PDBs, so Fingar invented a system. Though his objective was to end CIA dominance, he brought in one of the agency's analysts, Steve Kaplan, to be his deputy. He decided not to move PDB machinery out of Langley, as it would make no sense to duplicate the CIA's in-house facilities elsewhere. Instead, he made it a requirement that CIA PDB documents should be externally scrutinized by experts in other branches of the intelligence community. His idea was reminiscent of the 1970s A Team–B Team competitive estimating system, except that it would be more cooperative and harmonious. Fingar's non-CIA experts would not work from inside Langley, where they might be in danger of going native. Rather, they would communicate via specially encrypted emails. No longer would other members of the intelligence community find out too late that the CIA was getting it wrong.[5]

Fingar took on not only the task of reforming the PDBs, where immediacy was the requirement, but also the challenge of sharpening medium- and long-term estimates. He later wrote of "a self-imposed need to improve the quality of NIEs," the National Intelligence

Estimates that guided US foreign and security policy. Like the philosopher Willmoore Kendall, who had warned in the 1940s that the CIA should resist its "compulsive preoccupation with *prediction*," Fingar favored less attention to "problems and perils" and greater concentration on "opportunities to shape events." Several senior CIA and National Intelligence Council analysts agreed that the system was becoming increasingly weighted toward short-term PDBs, and away from "deeper research." A cynic might say that Fingar and his colleagues were buying into a system that provided insurance against censure for predictive failure, but a realist could be forgiven the conclusion that Fingar identified what intelligence analysts could do best.[6]

Fingar would write enthusiastically about changes taking place under the aegis of the Bush administration. One of these changes was the privatization of intelligence services. Awash with money in the wake of 9/11 and needing to hire fast, the CIA turned to the private sector, the "intelligence-industrial complex." The agency's action was consistent with trends in the military and with the laissez-faire ideology of the Bush administration. Major companies like Booz Allen Hamilton exploited the new opportunities. So did smaller enterprises like Abraxas Corporation of McLean, Virginia. For example, Abraxas supplied Mary Nayak to advise the CIA's internal review group on 9/11. Previously, she had run the CIA's South Asia intelligence unit. Abraxas had raided Langley for a human asset and then sold it back.

Nayak was one of many who remained in what had now become the intelligence industry, having exchanged the blue badge of a federal employee for the green badge of the profit-making sector. The process was known as "bidding back." Some of the rehiring even took place in the CIA cafeteria. The journalist James Bamford observed that when the CIA resumed employment of specialists it had trained at taxpayers' expense, it was sometimes at double the former rate of pay. Another trend was for ex-CIA persons to join private firms in the intelligence technology field, facilitating those companies' access, via subcontracts, to funds dedicated to national security. Among the several who were in post at the agency at the time of 9/11 and subsequently joined the private sector were CIA director Tenet and his counterterrorism chief, Cofer Black.[7]

By 2007, the number of private contractors working for the federal government on security matters and covert actions had reached 37,000. A leak from the office of the director of national intelligence revealed

that around 70 percent of the nation's intelligence budget went to the
private sector. At the CIA station in Islamabad, the privately paid outnumbered those on government payrolls by two to one. Legislators
took note of what was happening. In 2009 a story surfaced about how
in 2004 the CIA subcontracted assassinations of al-Qaeda personnel to
a private firm, Blackwater/Xe. Senate Intelligence Committee chair-
woman Dianne Feinstein (Democrat, California) censured the agency
for not keeping Congress informed about the operation.[8]

Fingar welcomed the consequences of privatization for the world of
the analysts. Outsourcing produced more data, and thus more work for
the analysts. Fingar noted with wry enthusiasm that there were now
"literally billions of times more 'dots' to be examined." On top of managing these billions, there was the challenge of reviewing and ranking
"9,100 cells in the matrix created by arraying roughly 280 international
actors against thirty-two intelligence topics," which yielded 2,300 priority issues. Fingar set out to prioritize the priorities and to choose an
agenda.[9]

Fingar's team looked beyond immediate crises and produced an
unclassified review called *Global Trends 2025: A Transformed World* (2008).
It identified global shifts and looked beyond the rise of the "BRIC"
nations—Brazil, Russia, India, and China—to identify new giants in
the making—Indonesia, Turkey, "and, possibly, Iran." *Global Trends 2008*
included a forecast of what might happen to global health and international relations, were an Asian-based pandemic to spread rapidly
through the world. Climate change, another threat to the nation's
long-term security, was another of Fingar's concerns. Not unduly
committed to modesty, Fingar stated, "I should probably take it as a
badge of achievement that members of Congress began to press for an
NIE on global climate change in late 2006 and early 2007."[10]

One of the problems confronting US policymakers ever since the
1950s had been the proliferation of nuclear weapons, especially in
unstable or unfriendly nations. The WMD threat from Iraq had proved
to be a fantasy, but a new storm soon gathered about the capabilities of
Iran. Secretary of Defense Rumsfeld claimed in 2003 that Iran had a
"very active program" aimed at the production of nuclear weapons.[11]
While Rumsfeld is routinely described as an administration hawk, subsequent revelations suggest he may have been correct in making his
allegation. In 2015–16 the Iranians would archive their reports and
CDs documenting their historical nuclear program. Israeli intelligence

procured about 20 percent of this archive. According to their inter-pretation of selectively released portions of the fragmentary evidence, the Iranians had been preparing to conduct underground nuclear tests in the years 1999–2003 and retained some capabilities after that.[12]

Rumsfeld and his fellow hawks issued remarks about the desirability of regime change in Tehran. By 2007, there was growing speculation about US military intervention. It was at this point that Fingar's team produced a national intelligence estimate that was a slap in the face for the hawks. It stated that Iran had ended its nuclear program in 2003. The estimating team sustained their conclusion by citing telephone intercepts where Iranian hard-liners complained about the decision to cease and desist. True to that policy, Iran had not restarted its program since.[13]

There was a furious response from conservatives such as John Bolton, recently the US ambassador to the United Nations, who pointed an accusing finger at "refugees from the State Department," a reference to Fingar and the supposedly soft diplomats of Foggy Bottom. *The Wall Street Journal* dismissed the Iran estimate as a concoction by Fingar and other "hyper-partisan anti-Bush officials." Peter Hoekstra, the Michigan Republican who chaired the House Permanent Select Committee on Intelligence, denounced the estimate as a "piece of trash." The intelligence veteran and historian Richard Immerman deplored the reactions of the critics. But for the hostile reactions, he argued, the accuracy and courage of the 2007 Iran estimate would have restored the reputation of the intelligence community.[14]

There was some speculation that Bush had known for some time that Iran had bowed to international opinion, and welcomed the esti-mate as giving both the opportunity to avoid another foreign war and the chance to leash the troublesome hawkish wing of the Republican Party. Although Bush remained reticent on the point, it was he who had authorized the release of the classified Iran estimate on 3 December 2007. This suggests that he wanted to harness its persuasive power.[15]

Be that as it may, Bush continued with politically calculated bluster against Iran. He warned of "nuclear holocaust" and "World War III" if Iran did not desist from what it was not doing. In 2008, a court case in Switzerland suggested that the CIA was plotting to channel defective components to both Iran and Libya in an effort to sabotage their alleged nuclear programs, and in future years Israel's Mossad was cred-ited with assassinating a number of Iranian atomic scientists.[16]

But war had been averted. The journalist Maureen Dowd articulated a widely expressed sentiment when she credited the crucial 2007
estimate to "Tom Fingar, a former State Department intelligence
officer who [had been] smart and brave enough to object to the
cooked-up intelligence on Iraqi WMD." For all that Fingar may have
aimed at theoretical adjustment and administrative reform, at the end
of the day he had, in the words of a much-loved cliché, spoken truth
to power. Thomas Fingar was able to stay at his post until 1 December
2008, embodying a singular rebuke to the way in which certain CIA
leaders had conducted themselves in the recent past.[17]

In spite of such success, the leadership of the intelligence community had a reputation for being "discontinuous and ineffective" in
Bush's second term, 2005–9.[18] Neither John Negroponte nor his successor as director of national intelligence, John McConnell, warmed to
his post, and neither director asserted his authority over the notoriously recalcitrant members of the community over which he nomin-
ally presided.

Porter J. Goss, who took over from Tenet as CIA director in 2004,
had an unenviable task in the wake of the 9/11 and WMD intelligence
debacles, and was then emasculated by the Reform Act, which stripped
him of the authority over the rest of the intelligence community that
he had inherited. He became the butt of ridicule. He was accused of
being too rich, too loyal to President Bush, too amiable, too much
addicted to drink, too hostile to intelligence reform, and too loyal to
his friends—he imported cronies to work at the CIA and, according to
his critics, these "Gosslings" knew little of intelligence and caused
demoralization at the agency.[19]

As in the 1990s, there was a too rapid turnover in CIA leadership.
Between Goss's resignation (effective May 2006) and March 2013, the
CIA's personnel experienced three more directors and an acting director, which gave an average length of tenure of under 1.8 years.
Perhaps it was indicative of their feelings of insecurity that Goss's successors resorted to populist tropes. Both Michael Hayden (2006–9) and
Leon Panetta (2009–11) emphasized their Catholic faiths. In doing so
they squared themselves with powerful evangelical trends in contemporary politics, but also made the statement that they were not members of the Protestant elite. Both brandished the credential of having
risen from "ethnic" neighborhoods. Hayden saw himself as embodying
the "blue-collar ethic" of western Pennsylvania. Panetta was proud of

being the son of a "peasant." He made a point of distancing himself from the "Ivy League spies of the early Cold War." Hayden and Panetta claimed virtues that were meant to distance them from the elitists like Yale graduate Goss in social terms and, they hoped, in degrees of competence.[20]

The foibles and turf wars of intelligence leaders may have contributed more to the appearance of ineptitude than to its reality. However, appearance, respect, and standing are intertwined. There was also a further threat to the standing of both the CIA and the new intelligence uber-directorship. This was to do not with culture wars but with the actual fighting wars favored by the champions of military solutions.

The most weighty of these was Donald Rumsfeld, who from 2001 to 2006 headed the Department of Defense for the second time in his career. President Ford had appointed him to that position in 1975, in the same cabinet reshuffle that put George H. W. Bush in charge of the CIA with the object of pushing back against congressional reform. Rumsfeld believed in an aggressive foreign policy based on worst-case scenarios. In 1998, he had chaired a commission on the ballistic missile threat to the United States. Its prime finding chastised the intelligence community for underestimating the threat posed by Iraq, Iran, and North Korea.[21] He adhered to the same patriotic pessimism in the wake of 9/11, when he famously declared:

There are known knowns. There are things we know we know. We also know there are known unknowns. That is to say, we know there are some things we do not know. But there are also unknown unknowns, the ones we don't know we don't know.

In the words of a British newspaper, he was "roundly mocked" for this gnomic wisdom. But it aptly summed up the rationale for Rumsfeld's shoot-first-just-in-case policy.[22]

Rumsfeld was just as pugnacious in his approach to domestic bureaucracy. He shared the military's concerns about the CIA's lack of combat professionalism. His undersecretary of defense, Stephen Cambone, found fault with the fact that the CIA housed both analysts and operators, which gave rise to the suspicion that the agency was grading its own performance. Rumsfeld wanted his department to expand into the realms of both espionage and paramilitary operations. He had at his disposal a specialist combat unit dating back to 1981 and known

as the Joint Special Operations Command. In the past it had performed paramilitary tasks, some of them in secret and others known to
the public. For example, its 1985 hostage rescue mission in Entebbe,
Uganda, following the hijacking of TWA Flight 847 from Athens furnished the inspiration for the 1986 movie *The Delta Force*.

Perhaps not appreciating that the Pentagon would be more constrained by the Geneva Convention than the CIA, Rumsfeld fancied
his unit could operate more widely than the civilian agency, moving
beyond the CIA with, as he believed, its hidebound regard for moral
niceties and potential congressional censure. There was some speculation that his proposed unit would not be bound, as the CIA was, by
executive orders still in force signed by Presidents Ford, Carter, and
Reagan, which prohibited assassination. The legal logic behind the new
assassination policy was that it was all right to kill people in foreign
countries without the permission of those countries' governments
because by harboring terrorists a nation surrendered its sovereignty.[23]

Rumsfeld sent his own small intelligence teams to US embassies
abroad, sometimes without the knowledge of the CIA. He also prevailed on Congress to enact a provision for a new post, undersecretary
of defense for intelligence. As things stood just after 9/11, the Pentagon
received 80 percent of the total US intelligence budget of $30 billion,
but the director of the CIA tasked the components of the Defense
Department, such as the National Security Agency, on how to spend
that money. Rumsfeld was aiming at a power grab. The incumbent of
his new undersecretary for intelligence post would later become
imbedded in the post-2004 Directorate for National Intelligence, as
well as being a weighty Pentagon figure. However, Congress, wary of
the increasing militarization of intelligence at a time when America
was being sucked into ever more wars, did stipulate that he or she
should be drawn from civilian life.[24]

Rumsfeld had more than the CIA in his sights. He targeted the new
Office for National Intelligence and had some success in undermining
it. Director of National Intelligence Negroponte was, as one of his
successors put it, "bureaucratically outgunned by Rumsfeld and the
large Office of the Secretary of Defense." When former CIA director
Robert Gates took up his post as Rumsfeld's successor in 2007, one of
his first actions was to declare his intention of mending fences with the
intelligence community, the CIA included. By then, Rumsfeld had
contributed substantially to the military tenor of US foreign policy.[25]

The US intelligence community underwent significant changes in the two administrations of President George W. Bush. The Intelligence Reform Act of 2004 helped to refresh the intelligence estimating process. Care should be taken not to exaggerate the marginalization of the CIA. In the words of a later chairman of the National Security Council, "We still looked to the CIA as our primary source of analysis."[26] But the agency no longer had the last say in producing the President's Daily Briefs and national intelligence estimates. Two other major characteristics of the Bush years, privatization and militarization, were informal developments that occurred independently of the 2004 legislation. They too, however, would color the legacy of the Bush reforms.

15

Battling al-Qaeda with Assassination and Torture

In the administration of George W. Bush (2001–9), the CIA resorted to torture and assassination, both pursued in a routine manner. It was a reaction to 9/11. It also sprang from the fact that America got bogged down in the intractable politics of Iran, Iraq, Pakistan, and Afghanistan—spawning grounds, all, of international terrorism.

In his inaugural address, President Bush promised to deliver democracy for the people of Iraq. But the destruction of that country's Ba'athist bureaucracy in the wake of its conquest in 2003 unleashed ungovernable forces. A minority Sunni Muslim government tried to hold out against the Shia Muslim two-thirds of the population and its Kurdish allies. Elections took place in 2005 and 2006 but failed to prevent sectarian bloodletting. The American policing operation by this time involved 130,000 US troops and a budget of $440 billion against the background of hemorrhaging public support back home. What was also very disturbing was the fact that all this chaos and bitterness formed an ideal breeding ground for al-Qaeda.

The same could be said not only of Afghanistan, where US ground troops fought to protect an anti-Taliban regime, but also of Pakistan. For although the Islamabad government of President Pervez Musharraf was relatively stable, it was an open secret that a section of Pakistan's inter-services intelligence agency known as "Directorate S" lent succor to al-Qaeda and to the Taliban who sheltered al-Qaeda, especially in refuges situated in the vertiginous borderlands adjacent to Afghanistan.

There was no straight military solution to such circumstances. The CIA found itself at the forefront of the campaign to deal with a far-flung panoply of terrorism. Immediately after the 9/11 attacks, the

Bush administration had launched Operation Greystone. Its mission was to hunt down bin Laden and other al-Qaeda leaders, and thus calm the nation's fears that even more horrendous events might take place, for example, attacks deploying nuclear radiation or anthrax. The president authorized the CIA to use a range of measures that included the isolation of captured al-Qaeda suspects in "black sites" or secret interrogation camps. Lines of command regarding treatment of prisoners were opaque. Military authorizations came directly from within the armed forces, whereas the CIA was supposed to receive congressional as well as presidential sanction. Responsibility for actions taken thus became blurred, "a mare's nest" for potential reformers in the assessment of one intelligence official.[1]

There was never any direct presidential order to torture prisoners, and there is some evidence that the president might have been protected from future condemnations by being kept to a degree in the dark. However, in 2007 President Bush gave a different impression when he formalized his policy in Executive Order 13440. While the order banned torture and murder, it justified rendition and detention on the grounds that al-Qaeda and the Taliban were illegal terrorist groups not affiliated to any state. It claimed that this gave the United States an exemption from full compliance with the Third Geneva Convention (adopted in 1929, revised in 1949, and ratified by the United States), which banned the torture of prisoners of war and the transfer of prisoners to non-signatory states.[2]

In Afghanistan the CIA failed to find and capture Osama bin Laden after searches based on the assumption that the Taliban were sheltering him there. It did, however, enjoy considerable early success. The agency deployed its own officers and also supplied a "sheep dip" facility: US Navy special combat Sea, Air, and Land Teams (SEALs) were temporarily designated CIA personnel, which allowed them, for reasons that would be stated in EO 13440, to disregard Geneva Convention regulations on the conduct of warfare. The CIA identified Taliban targets for the SEALs and for US bombardment. It contributed to the capture of the Taliban-held Afghan capital, Kabul, in November 2001.

It is reasonable to say that by going on the offensive overseas the agency weakened the terrorist threat to the homeland. Its work in the course of President Bush's first term in office was so effective that it contributed to the decision not to implement the 9/11 Commission's recommendation that the CIA should surrender to the Defense

Department and its Special Operations Command the "lead responsibility for directing and executing paramilitary operations." In refraining from adopting that recommendation, the Bush administration may also have been reacting against Rumsfeld's bellicosity. For whatever reason, the CIA remained in charge.[3]

In its war on terrorism, the CIA used euphemistic vocabulary suspended somewhere on the spectrum between George Orwell's "newspeak" and Donald Trump's "fake news." Its "rendition" was the practice of kidnapping a suspect in one country and then "rendering," that is, secretly and illegally transporting, that person to a "black site" (illegal prison) in another country. There, the suspect would be subjected to "enhanced interrogation" (torture). In addition to the clandestine black sites, the CIA was able to use Guantánamo Bay, an American naval base in southeastern Cuba that Havana had ceded to the United States under duress in 1903. Starting in January 2002, the CIA, with the cooperation of other agencies and with the secret collusion of several foreign governments, imprisoned 779 men in Guantánamo.

In December 2002 Tenet stated that the CIA had killed or captured more than a third of al-Qaeda's leaders. For example, in Yemen it had facilitated a missile strike that ended the life of Qaed Senyan al-Harthi, who was suspected of involvement in the bombing of the USS *Cole* that killed seventeen US sailors in the port of Aden in 2000. In February of the following year, the *Washington Times* reported that unspecified "authorities" had arrested twenty-eight al-Qaeda suspects in Naples, Italy.

Further north in Milan, a man masquerading as a police officer interrupted Hassan Mustafa Osama Nasr as he walked to his mosque. The CIA knew him to be a radical cleric who went by the name of Abu Omar. A few years on, Milan prosecutors identified twenty-six Americans and five Italian intelligence officers who, they alleged, had cooperated in the abduction of this one cleric. The CIA "renditioned" Abu Omar a distance of 1,600 miles to a notorious prison in his native Egypt, where he experienced surrogate enhanced interrogation, or "outsourced torture," to use the expression favored by the *Chicago Tribune*. Abu Omar filed a deposition saying that he was beaten, hung upside down, and subjected to electric shocks until he foamed at the mouth and became incontinent.[4]

In June 2002 *Newsday* drew attention to a novel aspect of the war on terror. The New York suburban newspaper told the tale of "the

RQ-1 Predator, a remote-controlled aircraft that relays video images from the battlefield to commanders in real-time," and of the Predator's deployment by the CIA's counterterrorism center in conjunction with the US Air Force. "The bright minds at the CIA," the paper's reporter Nathan Hodge wrote, "have tooled up the Predator with anti-tank rockets to assassinate al-Qaida leaders."

Hodge gave an example of Predators in action. Recently, one of them had targeted Gulbuddin Hekmatyar, previously a warlord financed by the CIA to fight Soviet occupiers of his homeland, Afghanistan. Now, given the shift of the tides, Hekmatyar was resisting the US presence in his land and was deemed to be a terrorist. The rocket attack occurred just outside Kabul. Hekmatyar survived, but several bystanders died, becoming, in the jargon of war, "collateral damage."[5]

In hawkish, conservative circles there had for some time been a growing unease about the 1970s ban on assassination. Although President Reagan reiterated the ban, CIA director Casey and his coterie favored a reconsideration. Detecting what was in the wind, in 1984 former diplomat George Ball urged the Reagan administration to examine the roots of terrorism, such as Palestinian grievances, and not to overreact and "embrace international lynch law and thus reduce our nation's conduct to the squalid level of terrorists." In more recent times, a senior intelligence official in the Obama administration noted the futility of assassination: "The terrorist groups were fairly quick to replace leaders" and "The [drone] strikes seemed to play into the terrorists' narrative; we were cowardly fighting from afar."[6]

Conservatives were deaf to such pleas, and there was extensive consideration of the assassination issue in the 1990s. One prestigious law review called for the executive order banning assassination to be refined, insisting that "Policymakers have a moral responsibility to consider it." Another voiced a view that was increasingly heard in conservative circles: "A unilateral declaration against ... assassination *increases* the risk of being victimized by ... an adversary."[7]

The day after 9/11, the UN Security Council called for all nations to cooperate to bring the terrorists to justice. It proclaimed, "All those responsible for aiding, supporting or harboring the perpetrators will be held accountable." American policymakers took it to be a justification for drone strikes in nations without the acquiescence of their governments. So the UN text was published alongside a congressional

"authorization for the use of military force against terrorists."
Californian Democrat Congresswoman Barbara Lee, the only legislator to vote against the resolution, denounced it a "blank check" for any
action. The wording of the resolution empowered the president only,
in that sense seeming to confirm that the CIA and its military partners
took their orders from the White House, and only from the White
House.[8]

So it came to pass that a sensor operator in the Creech Air Force
Base in the Nevada desert could with one twitch of her joystick
destroy life, also known as a "high-value target," in an eerily familiar
Afghan landscape 7,500 miles away. George Brant captured the scene
in his award-winning play *Grounded*. His operator was the mother of a
little girl:

Lingering over the dead. A mound of the dead. Our dead. They were ambushed
and I am to linger over their bodies. I do. With no idea of who they are or how
they got here. A mound of our grey. Our boys in grey. Please let me find the
guilty who did this the military age males who did this.

The girl Her face She stops running and I see it Her face I see it clearly I can
see her It's Sam It's not his daughter it's mine.[9]

The issue of torture gained political prominence and was a cloud
hanging over Bush's second term. First Amnesty International in 2003
and then CBS News in April 2004 exposed what they presented as
human rights violations in Iraq's Abu Ghraib prison. There were
accusations that US interrogators had subjected captives to sleep
deprivation, exposure to extreme heat, bright lights, and loud sounds,
inadequate toilet facilities, and sexual humiliation, including sodomy.
CBS deployed graphic photographs to sustain its charges. Local commander Brigadier General Janis Karpinski mounted a defense of prison
conditions and argued that terrorist suspects were not covered by the
Geneva Conventions as those conventions protected only nation-state
soldiers. But the Abu Ghraib disclosures shocked public opinion, and
although it was an Army scandal, the revelations sensitized Americans
to forthcoming accusations against the CIA.[10]

To illustrate those accusations, there is no better example than the
kidnapping and treatment of Khalid Sheikh Mohammed, a Pakistani
with over fifty aliases, usually referred to as KSM. With help from
Pakistan's interservices intelligence, the CIA captured KSM in
Rawalpindi on 1 March 2003. The CIA rendered him to black sites,

first in Afghanistan and then in Poland. Finally, the agency transferred him to Guantánamo Bay. Under enhanced interrogation, he admitted in 2007 to having masterminded the 9/11 attacks inspired and outlined in principle by Osama bin Laden in a meeting in 1996, as well as to several other further atrocities. His confessions came after he was placed in stress positions, subjected to sleep deprivation, and forced to suffer rectal hydration. He was also subjected to 183 applications of "waterboarding," a form of enhanced interrogation used five hundred years previously by the Spanish Inquisition, at a time when priests did not mince words and called it "torture."[11] It is a procedure whereby water is poured over a cloth on the victim's face to produce a sensation of drowning and, if not interrupted, it leads to brain damage and death. Further, it was made known to KSM that his children were being subjected to psychological torture.

The manner of KSM's treatment and his claims to have confessed under duress were the subject of numerous congressional hearings. Waterboarding caused public consternation amongst the nation's allies too. For example, a British Army officer walked out of "a working dinner with our US opposite numbers (probably not the CIA)" when "told by one of them that, 'Technically, you can't call it waterboarding if you're using petrol'."[12]

Setting aside the morality of the subject, the efficacy of torture has long been debated, and the debate revived when details of the CIA's methods inevitably leaked. Doubters say that the victim will say anything in order to stop the pain, so his or her evidence is unreliable. Another argument runs that when a victim realizes that nothing he or she says will stop the torture, they offer no more information because there is no point in doing so. However, enhanced interrogation had its defenders. It was supposed to be scientific, as two Air Force psychologists had devised the methodology. Drs. James Mitchell and Bruce Jessen had designed it through reverse reasoning, as they had previously advised US personnel on how to keep secrets when captive and under interrogation.

At Guantánamo, Mitchell and Jessen were present as observers and protested that CIA interrogators did not know when to stop. But the champions of torture called them "pussies." They pressed on and remained devoted to their craft. They did not lack supporters. In 2004, a voyeuristic account of Abu Ghraib practices appeared in the *Christian Science Monitor* claiming that the interrogator needed the "inspiration of an artist." The CIA's Michael Morell, one of President Bush's daily briefers, offered an impassioned defense of enhanced interrogation. He made the "moral equivalence" point that KSM and company were guilty of horrific crimes and were undeserving of sympathy. In terms of efficacy, he insisted that the torture of KSM had yielded a "treasure trove" of information that could be used to crush al-Qaeda.[13]

Every good scandal needs a cover-up. In the case of enhanced interrogation, Jose Rodriguez provided it. The cover-up occurred because CIA operatives had made self-incriminating videotape recordings of some of the interrogations that involved waterboarding and similar harsh practices. They deployed the video cameras, in part, to create a record and an archive that could be consulted for visual clues. Another reason for archiving arose from the questioning of suspected senior al-Qaeda operative Abu Zubaydah in the spring of 2002. Zubaydah had been wounded in the gunfight that resulted in his capture. He was in ill health, and the videos that recorded his every moment 24/7 were intended to show that his death, if it occurred, was not the fault of the CIA.[14]

On 8 November 2005 Rodriguez, who was head of the CIA's National Clandestine Service, ordered the destruction of the tapes. He consulted the CIA's lawyers before doing so. On 9 November the tapes were destroyed, the legal justification being that the CIA's inspector general "no longer required" to see them and that "cable traffic accurately documented" the "activities" in question.[15]

Textual records would be much less vivid than video tapes. Rodriquez was canny enough not to consult his superiors in the agency, for he realized that the destruction of the tapes would lead to charges of a cover-up and knew that officials from the president down would not want to be associated with that. He gave as his real reason for the destruction of the tapes his fear that they might be made public and create a furor just like the visual images of Abu Ghraib. One historian has paraphrased his concern with the words that their release would "inflame riots around the globe." Furthermore, their publication would expose the identities of the CIA officers involved, which would make them no longer able to undertake clandestine work and possibly endanger their lives.[16]

At the local level, Gina Cheri Haspel took care of the destruction of the tapes, cabling that an "industrial-strength" shredder should be used. An "Air Force brat" who had studied at an English high school and the

University of Kentucky, Haspel had acquired undercover experience in Ethiopia, Turkey, and Azerbaijan. She impressed Rodriguez as a promising CIA officer. He put her in charge of the black site in Thailand to which Zubaydah had been rendered. She was present at some waterboarding episodes, but did not personally participate in the torture of the wounded Zubaydah, with whom she had a "good" relationship—he called her his "Emira," Arabic for "princess."[17]

Gina Haspel's complicity in the destruction of the tapes would be significant in the politics of the years to come. As an aside, it is also worth noting that her ascent of the promotion ladder was an indication of how the CIA was trying to broaden its recruitment practices by hiring more women.

So did women make a difference at the agency? Did female spies change the clandestine profession, just as women at home and abroad had a distinctive impact on foreign policy? *New York Times* journalist Deborah Solomon toyed with the issue in her 2005 interview with Melissa Mahle, another woman who had served the CIA in the Middle East:

Q: I still don't understand why the C.I.A. would have put a California-bred blonde like yourself in the middle of East Jerusalem to monitor Palestinian extremist groups. Didn't you stick out?
A. When I wanted to be in the clandestine mode, I would use disguise. It's easy in the Mideast, because you put on a veil and the traditional black tent dress, and you look like everyone else...
Q. Would you go as far as having a love affair to get a source to spill his secrets?
A. That's against the guidelines of the agency. That's James Bond. Actually, James Bond would have been fired on the first day if he worked for the C.I.A....
Q. Did you find it hard to be a female spook in the masculine realm of espionage?
A. No. The same skills you learned as a little girl can be modified and used when you are working overseas as a spy. A C.I.A. officer has to be a good manipulator.[18]

The Bush administration may have aimed at liberalization in some respects, but it stuck to its policy of treating suspects very roughly, even when that practice came under fire. Yet there was no lack of policy alternatives. One was to work for a solution to that prime generator of Arab and Muslim anti-Americanism, the Israeli–Palestinian conflict. CIA director Tenet had already participated in one such effort on behalf of the Clinton administration, and he continued to see the

conflict as "a root cause of the global terrorism that plagues the world."
In 2003 the CIA was entrusted with looking after the security side of
the efforts of the "Quartet"—the European Union, Russia, the US,
and the UN—to implement a "road map" to peace in the region.
Irresolute political support led to the collapse of that initiative.[19]

Another putative peace initiative was economic in character. There
was talk of a "Marshall Plan" for Iraq that would have been a salve for
the bitterness from which support for al-Qaeda flowed. In constant
2005 dollars, the United States Marshall Aid plan had invested $29.3
billion in war-damaged Germany between 1946 and 1952. There was a
comparable investment of $28.9 billion in Iraq in 2003–6. However, 38
percent of that went into security, and the chief commitment of the
United States remained to the use of force, whether overt or covert.[20]

Bush had to contend with a growing challenge from those described
by one authority as the "legalists."[21] This put the administration on the
back foot, and the question that became a matter for public debate was:
did President Bush authorize torture? The *New York Times* correspondent James Risen commented on the absence of a paper trail leading to
the Oval Office: "It was as if the interrogation policies were developed
in a presidential vacuum." But he did note Bush's comment when told
that Abu Zubaydah's painkillers made him so groggy that he could not
answer questions put to him by his interrogators: "Who authorized
putting him on pain medication?"[22]

Interviewed by journalists on her way to Berlin in December 2005,
Secretary of State Rice insisted, "The President made very clear from
the beginning that he doesn't condone torture."[23] However, the issue
would not go away. On 29 June 2006 the Supreme Court ruled that
the administration was bound by the Geneva Conventions on the
treatment of prisoners and on torture. Soon after that, Bush had to
admit to the existence of black sites. His response was to close them.
He arranged for the prisoners to be transferred to Guantánamo. By
this time, in the guarded words of a congressional research report, the
nation's senior spooks "acknowledged that the US Intelligence
Community confronts a major challenge in clarifying the roles and
responsibilities of various intelligence agencies with regard to clandestine activities."[24]

In February 2007 Rice had to engage in legal gymnastics in justify-
ing the administration's position to the Senate Armed Services
Committee. She referred to "inherently vague" terms in the Geneva

Convention and to the "U.S. reservation to the Convention Against Torture" which justified American legislation and action in regard to detainees.[25] Two months later, the House of Representatives held a hearing on the effect of US counterterrorism actions upon America's alliance with Western Europe. Congressman Bill Delahunt said that European sympathy for the United States had plunged since the 9/11 *Le Monde* headline that had declared, "Today, We Are All Americans." A Pew opinion poll conducted in 2006 had suggested that people in the UK, France, and Spain saw US policy in Iraq as a greater threat to world peace than Iran's putative nuclear program.

Some witnesses pushed back against the criticism. One of them noted that fourteen foreign countries had cooperated with the CIA, for example, by supplying "black sites" for interrogation. European countries had allowed no fewer than 1,245 CIA overflights, landings, and takeoffs. Michael Scheuer testified, confirming he had initiated renditions and stating it was "passing strange" that European governments and the European Union protested about the CIA's civil rights abuses while harboring terrorists and refusing to extradite them. Collectively, the hearings indicated that the CIA's actions were provoking no more than ambivalent responses from allied leaders but were alienating the European citizens who voted those leaders into office and in this way were embarrassing friendly governments.[26]

In the following year, a *New York Times* editorial denounced CIA interrogation practices as "horrifying," "unnecessary," and not listed in the US Army field manual. The CIA replied that the manual did "not exhaust the universe of lawful interrogation measures available to the republic to defend itself against hardened terrorists" and claimed that the CIA's actions were "fully consistent with the Geneva Conventions and with current United States law."[27]

In August 2008 the US Senate sought to prevent the president from modifying his executive orders in secret, a move that indicated significant distrust of the White House.[28] The administration remained unmoved. When the US Army revised its field manual to conform more closely with the Geneva Conventions on interrogation, both houses of Congress demanded that the CIA also confine its methods to the practices detailed in that same manual. President Bush vetoed that legislation on 8 March 2008.[29]

Meanwhile at Guantánamo, where 275 men remained in custody, preparations were under way for a military court to try KSM and five others charged with 9/11 offenses. With the approach of a presidential election, the media once more latched on to the political significance of an intelligence issue. Some scribes supported the administration. One journalist wrote that the Guantánamo "case could begin to fulfill a longtime goal of the Bush administration: establishing culpability for the terrorist attacks of 9/11."[30] By this time, however, it was too late to prevent the CIA's tactics from becoming politically toxic for the Republican Party. CIA director Michael V. Hayden (2006–9) seemed prepared to bend with opinion and to accept that there were legal questions to be answered about the interrogation methods used by his men and women. Bush and Vice President Dick Cheney rejected that view, but at a price. As another journalist put it, harsh interrogation had become "a stubborn political burden."[31]

According to the journalistic watch, the issue played a salient role in the presidential election of 2008: "For two years on the presidential trail," Mark Mazzetti and Scott Shane wrote, "Barack Obama rallied crowds with strongly worded critiques of the Bush administration's most controversial counterterrorism programs, from hiding terrorism suspects in secret Central Intelligence Agency jails to questioning them with methods he denounced as torture."[32]

In the event, intelligence scandals as such had only a minor impact on public opinion in the run-up to the 2008 election. The war on terror was beginning to take a back seat politically. According to Pew Research, potential voters were 5 percent less worried about terrorism in 2008 than in 2004, which perhaps favored the campaign of the Democratic candidate, Barack Obama. There was also a sagging in support for the long-drawn-out and still inconclusive war in Iraq. As ever, domestic issues loomed large. Hurricane Katrina had in August 2005 wrought havoc on the Gulf Coast and especially in New Orleans, and the Bush administration came under fire for its allegedly inadequate response to the needs of the citizens affected. Above all, the economy engaged the attention of the electorate, with 87 percent of the Pew-polled expressing concern. The global financial crisis of 2007–8 was bad news for the incumbent Republicans, associated as they were with big business. The housing market collapsed, and the bankruptcy of the investment bank Lehman Brothers in September 2008 seemed for an illusory moment to herald the end of capitalism. In addition to all this, the ethnic composition of the United States was changing in a manner that potentially disadvantaged the Republicans. Obama won the presidential election with only 43 percent of the white votes cast.[33]

Although such considerations dwarfed the issue of intelligence abuses, promises made on the campaign trail are meant to be kept. Obama had reaped at least some advantage by denouncing torture on the stump. Once elected on the slogan "Yes, we can," it was incumbent on him to introduce change.

16

Obama's CIA and the Death of bin Laden

Back in the 1970s, President Carter had at first aligned himself with the critics of CIA excesses. He caused bitterness when he oversaw a substantial reduction in the number of covert operational personnel. Then, toward the end of his term of office, his attitude changed. For in difficult times he had experienced the burdens of decision-making, the advantages of good information, and the convenience of a covert option. Senator Moynihan quipped, "Carter has now discovered that it is *his* CIA!"[1]

In January 2009 another liberal Democrat succeeded a Republican in the White House. Like Carter, President Barack Obama had criticized the CIA in his campaign for office. However, his approach to the agency was different as soon as he assumed power. It is true that he continued to tread the moral high ground. He proclaimed his devotion to the "rule of law" and banned the use of torture. He announced the arrival of a "New Era of Openness," and soon Congress would legislate anew against the overclassification of information. But Obama felt Republican patriotism breathing down his neck. He was keen to demonstrate his devotion to national security by utilizing the CIA. He felt the need to show toughness. As he later declared, a "new, liberal president couldn't afford to look soft on terrorism." While he had taken issue with President Bush for starting wars, they were "my wars now."[2]

Obama did not ask General Hayden to remain at the helm in Langley and instead appointed a new director, Leon E. Panetta (2009–11). This decision for discontinuity signaled potential partisanship and politicization. However, Panetta had a bipartisan record. The new director emphasized his humble Greek-American origins and ticked the anti-Ivy League box in a manner that played into a populist mood that could be found in both political parties.[3]

If Panetta declared he was keen to restore the CIA's "standing with the American public," it was not at the expense of his relationship with the White House and with the federal bureaucracy. According to one senior CIA official, Panetta enjoyed "direct access" to Obama in a way that stifled the ambitions of his bureaucratic rivals. He successfully defended his agency's remaining prerogatives. For example, when he retained for the CIA the right to appoint station chiefs in distant parts, it was a triumph against attempted encroachment by the Office of the Director of National Intelligence. The CIA held onto the privilege of preparing the President's Daily Brief. Each day, Obama would arrive for breakfast and open the leather binder left at his table. It contained a few top-secret pages that had been prepared overnight. Michelle Obama invented her own name for the brief—"The Death, Destruction, and Horrible Things Book." Its delivery each morning meant that the agency was regularly delivering to a highly literate chief executive the kind of briefing that was likely to engage his mind. From the beginning, the beast of Langley had become Obama's CIA.[4]

As one might expect in the lengthening aftermath of the 2004 Reform Act, other players remained in the field. Old rivalries with the military showed no signs of abating, and the newly assertive ascendant National Intelligence Council played a role. The analytical body by now had a staff of five hundred. Partisans of the National Intelligence Council claimed a legitimating lineage that stretched back to the research and analysis division of the OSS. It was a claim that signaled status ambitions and a tendency to belittle the CIA, even as the council relied on the agency for the bulk of its analyzed data. The president and the National Security Council showered the National Intelligence Council with tasks. According to Christopher Kojm, its chairman in the years 2009–14, the council produced "several hundred" two- to three-page reports each year in response to the "countless memos" it received from the Obama administration. The memos were evidence of a president with an inquiring mind, but some senior analysts lamented what they perceived to be short-termism, a preoccupation with immediate political problems to the detriment of more thoughtful national intelligence estimates that looked into the future.[5]

The council did address some medium- and long-term issues. Notably, as it had in its 2007 Iran estimate, it calmed nerves about certain perceived threats. For example, it concluded that China was unlikely to ally with Russia to challenge the global position of the United States. Russia's President Vladimir Putin was too proud to play second fiddle to a larger partner. Perhaps optimistically, the council offered the view that "China still cared enough about its relations with the United States to avoid needlessly irritating Washington."[6]

The CIA had to fight for its slice of the intelligence budget, and, in spite of President Obama's keen interest, that budget was in decline. For while Obama took responsibility for his wars and for his CIA, he was also committed to military reductions, including withdrawal from Afghanistan, and this had repercussions for the intelligence community.

Within the intelligence community itself there was a feeling in some quarters that the time was ripe for reductions in the now vast homeland security bureaucracy. One would not expect the leadership to express that view, but some rank-and-file members of the CIA and its sibling agencies sensed the demoralizing effect of overstaffing. Bridget Rose Nolan, who worked in the CIA's counterterrorism unit, trained the eye of a sociologist on the attitudes of her colleagues. She recorded those attitudes in her dissertation when she left to complete a Ph.D. at the University of Pennsylvania. There was little sympathy for the new behemoths, the National Counterterrorism Center (NCTC) and Department of Home Security (DHS), that had been superimposed on the intelligence community in the Bush years:

When I asked Jack, a 15-year CIA veteran, if things were better or worse since the creation of NCTC, he said this:

Jack: Uhh ...I'm not a fan of NCTC ...I think it's a mistake.

Bridget: So do you think the CIA director should still be the leader of the IC?

Jack: No doubt.

Chad, also of the CIA, noted "a lack of coordinated effort among the core agencies" regarding counterterrorism. He said, "DHS is off doing random stuff—half of us don't even know why they exist or what they do . . . There is a lot of duplication." Thirty-eight-year-old Pamela, formerly with the FBI but now an analyst in the Office of the Director of National Intelligence, declared of the NCTC:

I think it really doesn't have a function or it doesn't improve anything enough to justify its existence. That's so terrible because I work there! I think that a lot of the information and production that comes out of NCTC can be done or is done better by CIA and FBI.

Nolan concluded, "There is a general sense . . . that NCTC was almost a knee-jerk reaction to 9/11," and "a way for the government to treat the symptoms, but not the cause, of the perceived problem."[7]

The Obama administration reduced the nation's intelligence budget. This had climbed from $63.5 billion in 2007 to $80.1 billion in 2010, but it gradually declined to $66.8 billion by 2015. Expressed in terms of constant 2014 dollars, the reduction between 2010 and 2013 was 13.5 percent.[8] Obama did believe in federal spending, but he had other priorities. Soon after his inauguration he signed the American Recovery and Reinvestment Act, an economic antidote to the recent recession that cost $787 billion. His pet domestic project, the Patient Protection and Affordable Care Act ("Obamacare," March 2010) would within two years cost an estimated $1.76 trillion. By the summer of 2011, Republicans in Congress were worried that the government was heading for a "fiscal cliff," and might even default on its bond issue.

Regardless of his bias toward domestic spending, Obama did make an exception. He singled out the CIA to be a recipient of his largesse. In August 2013 the *Washington Post* obtained a 178-page summary of the "black budget," an itemized list of intelligence expenditures that was normally top secret. It revealed that the CIA was receiving $14.7 billion, far more than generally thought and more than any other member of the intelligence family. Surprisingly, the CIA's budget was equivalent to 150 percent of the budget of the National Security Agency with its costly computer expenditures.[9]

President Obama strongly supported the CIA's program of drone warfare. With Panetta and Obama's antiterrorist adviser John Brennan playing a leading role, the CIA engaged in "targeted killings" of suspected al-Qaeda and Taliban leaders. Brennan had been Obama's first choice as CIA director, but had at that time been too closely associated with some of the agency's past interrogation practices to be acceptable to the liberal coterie surrounding the nation's first African American president. He defended the drone program as legal, wise, and ethical. On the moral issue, he contended that drone attacks conformed to the principles of proportionality and of distinction of military from civilian targets. Also, they conformed to the principle of necessity. This meant that the CIA had the "authority" to kill al-Qaeda suspects, "just as we targeted enemy leaders in past conflicts, such as German and Japanese commanders during World War II."[10]

The drone program had three objectives: the "decapitation" of terrorist leadership, the denial of a "safe haven" in Pakistan to those terrorists, and through these means the protection of US and allied armed forces in Afghanistan. By killing his enemies instead of torturing them, Obama was able to distance himself from the opprobrium that had attached to the Bush administration, and the program had the further political advantage that there would be no US body bags.[11]

To begin with, Predator and Reaper drones operated over the tribal areas of Pakistan, where al-Qaeda and the local population came to dread the perpetual overhead menace. Thereafter, operations expanded. Camp Lemonnier in Djibouti in the Horn of Africa developed as the control point for a number of other bases. Drone strikes now took place against suspected militants in Yemen, Somalia, and Libya. The program was initially top secret, but in 2016 the administration confirmed that 473 strikes had taken place, a figure that dwarfed the 46 when Bush was president. The official figures indicated that between 2,372 and 2,581 combatants had been killed since January 2009, and 64 to 116 civilians. One intelligence official observed that the CIA had become a "killing machine," but the journalist David Ignatius greeted the development with the headline "CIA Back on its Feet."[12]

The liberal media supported Obama as the champion of social reform. Yet the drone killings stirred the conscience of some of his supporters. How could the winner of the Nobel Peace Prize be president at a time when the CIA engaged in activities that denied due process, assumed guilt without a fair trial, and accidentally killed innocent people?

There was skepticism about the official figures. A staff writer in *The Atlantic* magazine contended that the "absurdly low [casualty] figures cited by the Obama administration were lies." He offered an example. A whistleblower had passed a cache of documents to an online magazine called *The Intercept* about Operation Haymaker, a CIA undertaking in northeastern Afghanistan:

Between January 2012 and February 2013, US special operations airstrikes killed more than 200 people. Of those, only 35 were the intended targets.[13]

Undoubtedly, innocent people were killed, in part because missiles and bombs are rarely as accurate as their makers claim, and in part because the CIA received and believed false intelligence. Agency officers created composite identities of what terrorists might look like and how they might behave. They called them "signatures," an idea reminiscent of the controversial way in which domestic agencies used to profile citizens they considered to be suspect. If a "jihadist" had a suspect signature a Predator operative twitched her joystick and sent him to paradise.

Revenge killings further dented the reputation of the CIA. One example stemmed from the time when, in December 2009, a jihadist called Humam Khalil al-Balawi tricked a CIA team in Khost, Afghanistan, into believing he would lead them to Obama bin Laden. Inattentive guards waved the "informant" into Camp Chapman, along with his suicide bomb. Al-Balawi thereupon shouted "There is no god but God!" and blew himself up, together with a crowd of CIA personnel, including the base chief Jennifer Matthews, an original member of the bin Laden-hunting "coven."

Seemingly to avenge the death of the seven officers, the CIA intensified its drone strikes in the North Waziristan region of Pakistan. The "don't tread on me" message may have been clear, but the agency was open to the charges that it valued its own to an excessive degree and that it was engaging in lethal action for an irrational reason.[14]

President Obama and his team took a yet more perilous step when they decided to execute American citizens without trial. The first of these was Anwar al-Awlaki. Born in San Diego, Awlaki was a charismatic imam (religious leader) who preached a message of hate, first in the United States and then abroad. He was thought to have inspired terrorists' actions. For example, he was known to be in touch with US Army Major Nidal Hasan, who shot dead thirteen people in Fort Hood, Texas, in November 2009, as well as with Umar Farouk Abdulmutallab, the "underwear bomber," who on Christmas Day of the same year failed in his attempt to blow up Northwest Airlines Flight 253 on its journey from Schiphol airport in the Netherlands to Detroit.

The CIA caught up with al-Awlaki in Yemen and placed him on its "goodbye list." In the case of this American citizen, the decision to kill was an open process that received the blessing of the Department of Justice and caused al-Awlaki's father to appeal in vain to the American courts. On 30 September 2011 operators in Nevada drew an electronic bead on their target as he was chatting in the Al-Jawf region in Yemen. Moments later, missiles killed al-Awlaki along with some others, including another American citizen. Soon after, a further strike inadvertently ended the life of al-Awlaki's son, who was a noncombatant, as was Jude Mohammed, a North Carolina high-school dropout killed in Pakistan. That made four Americans killed without due process. Under pressure to justify the government's behavior, the US Department of Justice produced a "white paper" on the lawfulness or otherwise of killing a US citizen who was a leader or "associated force" with al-Qaeda. It concluded that lethal force could be applied where "capture is unfeasible."[15]

In the biting words of the Pulitzer Prize-winning columnist Maureen Dowd, President Obama "feels tough when he talks about targeted killings."[16] The president could also applaud other advantages to the program of assassination-by-drone. Al-Qaeda was hit hard, losing much of its leadership. Its followers began to drift into ISIS, a rival terrorist organization that aimed to establish a caliphate (Islamic theocracy) by force and was gaining ground in war-torn Iraq. Former Bin Laden Station head Michael Scheuer took an ultra-realist view of this and other sectarian warfare in the Middle East: "All our enemies are killing each other and it is not costing us a cent or a life."[17]

America's use of lethal force did not, however, halt the worldwide spread of terrorism. According to State Department estimates, there were 6,700 jihadist attacks in 2012. The figure for the following year was 9,700. The cumulative death toll inflicted by terrorist acts in those two years was 18,000, with a further 33,000 injured. Would those numbers have been even higher but for Obama's tough policy or did they occur, in part, because of it?[18]

According to its critics, the CIA concentrated too hard on drone warfare. Drone jobs were the cream of the cream, attracting 20 percent of the agency's analysts and offering the "targeters" a fast track to promotion. That meant, so the censure continued, that the agency lost sight of what was going on in the cybercafés and comedy clubs where young people flocked in cities like Cairo. According to the former CIA analyst Roger George, the CIA had "reverted to an organization that is more like a war-fighting agency than a twenty-first century knowledge-based think-tank…it is back to the future. The CIA is now a re-born Office of Strategic Services (OSS)."[19]

Such militarization explained, in the critics' view, why Washington was caught unawares by the pro-democracy uprisings of youthful pro-testers in Tunisia (December 2010), Egypt (January 2011), Syria (January 2011), and Libya (February 2011). For there had been a scarcity of CIA agents who would sit in Cairo's Tahrir Square listening to a local blog-ger as he sipped his sweet mint tea sitting on a plastic crate. One such blogger revealed the contagion of the hour to a British journalist, insisting that if "Tunisia, a small country we beat at football, could do it, so could Egypt, the mother of the world."[20]

If the CIA failed to anticipate the Arab Spring, it was not because it was inactive in the nations bordering the eastern Mediterranean. In 2008 it had discovered a convoluted plan to supply a nuclear bomb sourced in Pakistan to Libya's Muammar Gaddafi, a dictator strongly suspected of fomenting terrorist attacks. Deploying Byzantine methods of its own, the CIA had foiled the plot.[21]

In exoneration of the CIA's performance in other areas, it could be argued that the muddied waters of Arabian politics, together with poor receptivity in Washington, made it impossible for the agency to make a difference. Another pitfall opened when Libya collapsed into a state of civil war after Gaddafi had fallen in October 2011. Security became a thing of the past. On 11 and 12 September 2012, Ansar al-Sharia, a terrorist group aligned with al-Qaeda, attacked American installations in Benghazi, killing US Ambassador to Libya J. Christopher Stevens and two private CIA contractors. Accusers pointed the finger at CIA negligence, but a Senate Intelligence Committee report found that the agency had given warnings of an impending attack. The committee report noted, "The Intelligence Community does not collect intelli-gence about threats to our security in dangerous places so it can be ignored by senior policymakers."[22]

Further north, Russia had wedged its way into the Near East's power structures by supporting the Syrian dictator, Bashar al-Assad. CIA officers based in southern Turkey tried to support the democratic opposition that had flowered peacefully in Syria's version of the Arab Spring and then turned into an armed insurrection. President Obama authorized Operation Timber Sycamore with the intention of arming and training these democratic opponents of the Assad regime. In April 2013 Obama redoubled the effort by authorizing the establishment of a CIA training and supply base in Jordan. However, the CIA commissioned an internal review of its covert training programs, which suggested they were rarely effective. Moreover, pro-democracy fighters and their CIA backers found themselves fighting on the same side as ISIS, which succeeded for a while in taking over a sizable portion of Syrian territory.[23]

If the CIA's failure to predict the Arab Spring cast a shadow, it was soon brightened by the glare of publicity that attended the elimination of Osama bin Laden. According to the account given by the Obama administration, the Bush administration had in its later years down-played the bin Laden issue. CIA director Panetta said it had become no longer a "top priority," because Bush wanted to obscure his failure to find the Saudi terrorist. President Obama, according to his retrospect-ive account, took a different tack. He told Panetta in May 2009 that he wanted "to make the hunt for bin Laden a top priority." His logic, as he recalled it, was that, in neglecting bin Laden, "We'd fallen into a strategic trap—one that had elevated al-Qaeda's prestige, rationalized the Iraq invasion, alienated much of the Moslem world, and warped almost a decade of foreign policy."[24]

Obama was able to act because, after more than a decade of detect-ive work, the CIA discovered bin Laden's hiding place. It was a custom-built, highly secure compound in an affluent suburb of Abbottabad, a city in northern Pakistan. In his politically freighted account of what led to the terrorist's discovery, the CIA's deputy director Michael Morell asserted that one of the vital connecting dots came as the result of "enhanced interrogation"—an idea the 2012 thriller movie *Zero Dark Thirty* would popularize.[25]

The operation called Neptune Spear launched on 1 May 2011. Vice President Joe Biden had opposed the idea of sending in a specialist team. Secretary of State Hillary Clinton was cautiously supportive. Defense Secretary Gates wanted just to use a large bomb that would kill everyone in the compound. Such actions did not reliably produce identifiable remains, and President Obama was prepared to risk a more surgical operation. The appropriate congressional committee members approved his plan. After being "sheep-dipped" to make them tempor-ary CIA operatives, a team of US Navy SEALs took off in Afghanistan in Black Hawk helicopters. They flew into Pakistan without notifying its government. They descended from the dark night into the suspect compound, catching its inhabitants by surprise.

The next day, unidentified "senior administration officials" briefed the press:

Osama bin Laden is now no longer a threat to America.…He was responsible for killings thousands of innocent men and women not only on 9/11, but in the East Africa embassy bombing, the attack on the USS Cole, and many other acts of brutality.

The officials stated that (serially waterboarded) Khalid Sheikh Mohammed (KSM) had identified the courier who led the agency's detectives to bin Laden's lair.[26]

According to the officials, there was "no doubt that the death of Osama bin Laden marks the single greatest victory on the US-led campaign to disrupt, dismantle, and defeat al-Qaeda." An excited crowd outside the White House greeted the momentous news, shouting "USA, USA, CIA" (and not, as one historian waspishly observed, "DNI, DNI").[27]

The killing raised a moral issue. In reply to a question, the officials replied, "He did resist the assault force. And he was killed in a firefight." Obama recalled that the order given to the SEALs was that "They would apprehend or kill bin Laden." That recollection raised the possibility of arrest. In the event, a SEAL—Obama said he did not want to know his name—shot dead in cold blood a man in his nightclothes who did not demonstrably go to bed holding a weapon (though he may have done). There was no attempt to take permanent custody of bin Laden either alive, to stand trial, or dead. Back in Afghanistan, his captors transferred his body onto another aircraft, whose crew flew south and dropped it into the capacious waters of the Indian Ocean without disclosing the coordinates. There was a feeling that that was the best way to prevent the arch-terrorist from becoming a martyr. In the words of one intelligence officer, "There was a real concern that access to the US judicial system would give him a megaphone he did not deserve. It might also have proved a destabilizing event for a prolonged period of time."[28]

There were some doubts about the operation that killed bin Laden. According to one train of thought, bin Laden was a has-been by the time he was killed and no longer had more than an ideological influence within al-Qaeda. The Pentagon saw the necessity of arranging a special press briefing that offered videos captured in the compound and other evidence to show that the late bin Laden had still been actively planning terrorist atrocities.[29] There was also the question of whether it was worth offending Pakistan, which had been in some ways helpful in combating terrorism. But many, especially those sympathetic to

President Obama and his CIA, would have agreed with the crowd outside the White House and would applaud the verdict of the historian who described the assassination of Osama bin Laden as "the CIA's most daring and successful covert operation."[30]

The video of the president and his advisers crowding intently in front of a screen as they watched the triumphant drama unfold in real time became an iconic image of the Obama presidency. Obama had run for office as a candidate who would achieve more social justice at home. While he succeeded with Obamacare, a Republican majority in the Senate had frustrated a great part of his domestic agenda. He stood in danger of being remembered instead for an illiberal deed, the summary dispatch of Osama bin Laden.

Obama continued to have a proprietary relationship with the CIA. By common consent he was on good terms with its leaders, who commanded his ear to a greater extent than the directors of national intelligence superimposed by the 2004 legislation. The CIA, nevertheless, continued to be collegiate. Christopher Kojm, chairman of the National Intelligence Council from 2009 to 2014, recalled that in response to the National Security Council his team continued to produce both weighty national intelligence estimates and hundreds of shorter intelligence memoranda. The CIA contributed substantially: "Its drafting contribution to NIC products significantly exceeds the contribution of any other IC agency."[31]

Still basking in the glory of bin Laden's demise, Leon Panetta quit the directorship of the CIA. He succeeded Robert Gates as secretary of defense. To replace Panetta, the president chose David Petraeus. A four-star general, Petraeus had recommended and then commanded a US military surge in Iraq, creating a sufficiently persuasive illusion of peace in that troubled land to allow the withdrawal of US troops that public opinion craved. Though he shed his Army uniform on joining the CIA, he carried particular military priorities into the agency, priorities that chimed with Obama's devotion to withdrawing ground troops and hitting America's enemies with high-tech weaponry.

If you visit the Berlin museum devoted to the history of the notorious Stasi secret police, you will see the director's breakfast table laid out with military precision. Petraeus did not serve a police state, but his high military rank had accustomed him to being similarly served. He denied that he wanted his breakfast bananas to be sliced just so, but the circulation of such rumors signaled that the agency was uneasy at being overseen by a man whose every other sentence seemed to be an order.

What led to Petraeus's premature downfall in the fall of 2012 was, however, a different kind of human foible, his infatuation with a fellow officer who was writing his biography. Twenty years his junior, Paula Broadwell was a married woman and had two children, as did Petraeus. The general's indiscretion went further than that. The FBI opened an investigation when it received complaints that Broadwell was apparently harassing another woman whom she suspected of having engaged Petraeus's overgenerous affections. Delving into email accounts, the FBI discovered thousands of CIA documents that were in Petraeus's personal email account, not where they should have been—in his secure CIA account. It transpired that he had shared with Broadwell information on national security, including the identities of covert officers. Although there was no suggestion that Broadwell had passed the information to foreign agencies, her lover had to go. Petraeus was already under pressure, perhaps unfairly, because of the Benghazi fiasco two months earlier. Director of National Intelligence James Clapper told him to resign.[32]

Obama appointed John Brennan (2013–17) in place of Petraeus. The son of an Irish immigrant, Brennan had in his youth worn a diamond earring and voted for the presidential candidate Gus Hall, which made him probably the only CIA director ever to have cast a ballot for a communist. A former high-school athlete, he now walked with a cane after several bone fractures and a bout of cancer.[33] He learned about moral relativism at an early age. After his first degree at Fordham, he studied for an MA in government with a concentration on Middle Eastern Studies at the University of Texas at Austin. He also spent time at the University of Cairo, became fluent in Arabic, and developed a particular perspective on democracy and human rights. In his UT Austin MA thesis, submitted in May 1980, Brennan predicted that there might be a rebellion brewing against the regime of Egypt's President Anwar Sadat. He thought that one reason, amongst others, was the regime's "relaxed socialist policies," which were frustrating social mobility. But he argued that a rebellion in the name of human rights might cause dangerous political instability. Torture and censorship were arguably defensible in the furtherance of stability. He thought there could be "no objective answer" to the question "Can the human rights violations in Egypt be justified from a democratic perspective?"[34]

Brennan was a senior CIA executive when the enhanced interrogation policy came into effect and had overseen the drone strike program in Pakistan. This ultra-realist liberal was so close to Obama that military personnel reportedly referred to him as "deputy president."[35]

Portraying himself to be at heart an analyst in spite of his paramilitary record over the last dozen years, Brennan pledged in his confirmation hearings to remedy the intelligence failings that had been evident at the time of the Arab Spring uprisings. He signaled his commitment to diversity by appointing black civil rights veteran Vernon Jordan to head an investigative commission. It found that racial and ethnic minorities made up 24 percent of the CIA workforce, but under 11 percent of senior management. Brennan also lumped analysts and spies together in mission centers. Some of these centers dealt with specific regions, such as the Near East, which mirrored the arrangements in the Department of State. Others dealt with security threats such as counterterrorism or nuclear proliferation. Obama meanwhile announced that the CIA's Counterterrorism Center would cease to be in charge of drone operations; these would move over to the Defense Department, which already had its own drone strike program.[36]

The CIA would in future concentrate on spying, and on analysis. That signaled a cultural shift, as most of the large intake of CIA personnel after 9/11 had been engaged, as journalist Mark Mazzetti put it, in "man-hunting and killing." Yet in spite of the promised shake-up, Brennan retained the loyalty of his CIA colleagues.[37]

The Obama-Brennan partnership did have to confront significant problems in the time remaining for the twice-elected president. One of these was the rise of cyberattacks, the use of computers to spy on or disable enterprises reliant on information technology. The targets could be in the defense establishment or in civil society, and could be sinister—in the UK, for example, there would be an attack on the National Health Service in 2017.

Vladimir Putin's Russia was not the only offender here, but it was a prime culprit. Russia was reacting to pressure. Americans were directing a stream of criticism at Moscow's increasingly autocratic regime. With its fellow North Atlantic Treaty Organization members, the United States strengthened Russia's neighbors to help them withstand possible encroachments. To Moscow, the policy smacked of encirclement. It stirred up age-old insecurities within the walls of the Kremlin.

Outgunned in military terms, Russia resorted to the relatively risk-free retaliation strategy of cyberwarfare. In the communist era, Moscow had sought to reform or convert Americans, but now the aim was disruption bordering on destruction.

The House of Representatives held a hearing on the subject in March 2013. Those testifying mentioned the role of the CIA on just two occasions. It was a double warning to the agency. It had to be alert to the developing threat, and, in dealing with the cyber menace, it was being outflanked by other agencies such as the National Security Agency.[38]

The National Security Agency was also central to the Edward Snowden affair. This time the CIA was glad to be on the periphery, but there was still cause for concern. Born in 1983, the son of an FBI official and a Coast Guard rear admiral, Snowden was a talented computer analyst. He worked for the CIA after winning certification as a Microsoft systems engineer in 2005, and it was when he was on an agency assignment in Geneva in 2009 that he began to think of whistleblowing. He thereafter joined the Dell corporation, worked on subcontracts for the National Security Agency, and purloined extensive files of classified information.

Liberal- and left-leaning commentators had cheered the whistleblowers of the 1970s, but Snowden held conservative, libertarian, anti-statist principles. They related to a growing tendency in American politics epitomized by the Tea Party movement within the Republican Party and by conspiracy fears about the "Deep State," fears that had once been characteristic of the populist left. Snowden chose to unleash his revelations in May 2013, when a liberal president occupied the White House.

An admirer of earlier leakers like Daniel Ellsberg, Julian Assange, and Bradley Manning, Snowden fed the *Washington Post*, *The Guardian*, and *Der Spiegel* a stream of embarrassing secrets on his way to permanent exile in Russia. It transpired that the American public were under surveillance, to a shocking degree, by their own institutions, both public and private. America's eavesdroppers tapped social media. They also spied on allies like Angela Merkel and Dilma Rousseff, the leaders of Germany and Brazil respectively. Assange's Wikileaks organization had earlier released documents purporting to show that the CIA had collaborated with Israel, when in early 2009 that nation's agents insinuated a memory stick carrying the Stuxnet virus in a successful cyberattack on Iran's uranium enrichment plant in Natanz. Snowden's leaks further evidenced the close intelligence relations between the United States and Israel and how that disadvantaged the Palestinians. For security reasons, the US government did not discuss how many CIA agents on active duty might have been compromised by Snowden's indiscretions. But that issue, as well as the issue of whistleblowing generally, was a worry for the men and women of the CIA.

From the beginning of John Brennan's tenure at the CIA, the issue of torture posed difficulties. A 6,000-page Senate study of enhanced interrogation prepared at a cost of $40 million contained, in the words of one journalist, "blistering" and "incendiary" accusations about how the CIA had misled Congress. The report remained classified, but Mark Lowenthal, who had served in senior positions both in the CIA and on the staff of the House Intelligence Committee, declared that it was still a "potential minefield for John Brennan."[39]

Senator Dianne Feinstein, the California Democrat who chaired the Senate Intelligence Committee, backed the report's findings and appealed to the White House to publish the study's executive summary. Brennan was in a dilemma, as although Obama had outlawed the use of torture on his very first day in office, the CIA director had to defend his employees. In the event, he issued a detailed rebuttal of the charges in the report. The Intelligence Committee's senior Republican, Georgia's Saxby Chambliss, spoke up in Brennan's support.[40]

The dispute dragged on and intensified in 2014, when the CIA accused Feinstein's committee of, in effect, spying on the CIA. The agency charged that committee staffers had used their privileges to access files containing an internal CIA study of the interrogation program approved by former director Leon Panetta. The committee's bitter riposte was that the CIA must have spied on "DiFi" and her colleagues to discover that they had spied on the CIA and obtained the Panetta report.[41]

Published in December 2014 as Feinstein had wanted, the executive summary included twenty indictments, of which the following are a selection:

1. The "enhanced interrogation techniques" were ineffective.
3. They were "brutal," and that fact was withheld from policymakers.
6. The CIA "actively avoided or impeded congressional oversight of the program."

13. The interrogation program had been entirely outsourced by 2005, the two psychologists involved having formed a company that collected $180 million for its services, together with an indemnity agreement. The psychologists were unqualified for the work they undertook, being, for example, ignorant of Al-Qaeda and of the necessary languages.

20. The program "damaged the United States'" standing in the world.[42]

Former vice president Dick Cheney described the Senate report as "deeply flawed," and there was a heated debate as to whether waterboarding really amounted to torture. Feinstein remained firm in her conviction that it was. She declared that the issue was "really about American values and morals." Arizona's Senator John McCain, who had been Senator Obama's Republican opponent in the 2008 presidential election, spoke with the experience of having been tortured following his capture in the Vietnam War. He agreed with Feinstein and issued a remonstrative reminder that the United States had "for the most part authored" the international conventions banning the use of torture.[43]

Meanwhile, in real time, the civil war in Syria was an all-too-evident humanitarian tragedy. The Russian-supported dynastic regime of Bashar al-Assad proved to be just as resilient as Egypt in resisting the tide of democracy. Those who opposed the regime included Islamic State fundamentalists as well as democrats. The divergent factions fought each other, and Assad spread fear by using chemical weapons. Obama had authorized the CIA to train and equip the opposition, but was this wise? The agency commissioned an internal study to see how it had fared in similar enterprises in the past. It concluded that, with very few exceptions, clandestine interventions of that type had not succeeded.[44]

In clutching the CIA to his breast as his own special toy, President Obama perhaps had an optimistic image of the agency and what it could achieve. His relationship with and trust in the agency seemed to diminish and even negate the reforms of the Bush years. In the final pages of the first volume of his memoirs, the president revealingly described how he relaxed in the aftermath of the bin Laden triumph: "I spent time with John Brennan."[45]

17

Fake News Comes Home

Presidents Putin of Russia and Trump of the United States shared a dream. Echoing the halcyon days of the Gorbachev–Reagan relationship, they envisaged a new dawn of harmony between their two nations, to the benefit of both. Aware of Donald Trump's sympathetic view of Russia, Vladimir Putin instructed his country's disinformation services to increase the chances that Trump would be elected president in 2016. This deployment of "fake news"—to use one of Trump's favorite expressions—was a tried and tested, yet discredited secret service methodology. The CIA had used it in elections in Italy, Chile, and other countries, often provoking hostile reactions. The agency's director John Brennan knew that Russian deployment of such interference against the United States might well be counterproductive for Moscow. In private meetings, he warned the Russians that their tactics would "backfire."[1]

Against the predictions of the pundits, Donald Trump was elected president on 8 November 2016. He had campaigned on an "America First" platform with an anti-elitist rhetoric. Hillary Clinton's environmentalist campaign alienated the Democrats' traditional blue-collar support, with workers believing Trump's promise that their coal mines and their rust-belt steel mills could be opened once again. These were the major issues. Russian interference was a marginal factor, but it may have made a difference because it was a close election. And whatever the impact of the interference, it would indeed backfire, just as Brennan had predicted.

Knowledge of Putin's handiwork dawned on America because of revelations made possible by CIA counterintelligence work. Offended by the suggestion that he had not won the election fairly, President Trump therefore turned on the agency, portraying it as having been at the hub of an elitist conspiracy to frustrate the will of the people. His rhetoric failed to achieve its ultimate goal. Trump became the first chief executive to be impeached twice, and he failed to gain re-election in 2020.

The CIA staggered under the weight of these proceedings. Its loss of standing in the Trump White House was mirrored nationwide and on novel partisan lines. Opinion polls showed that the CIA had climbed in the public's esteem to a peak at the time of bin Laden's death and then declined. A poll for the *Wall Street Journal* identified a switch in partisan attitudes. Ever since the Reagan presidency, the agency had been the darling of Republican voters. But on the eve of Trump's inauguration it had a net approval rating of 32 percent amongst Democrats, but only 4 percent with Republicans, many of whom were evidently in sympathy with President Trump's contention that the CIA had conspired against him.[2]

Russian interference in American affairs was not new, but intensified in 2016. The CIA was able to keep track of the marked escalation because it had a mole in the Kremlin. The agency exfiltrated the mole when the media learned of Putin's plot to suborn an American election and about the source of that information, but by this time the mole had done their work. With the advantage of intensive surveillance of Russia assisted by its mole, the CIA had developed a good idea of what was going on. It was aware that DCLeaks.com and Guccifer 2.0., two Russian cyberwarfare entities, had in June 2016 released thousands of hacked emails from Hillary Clinton's Democratic campaign. Additionally, some 20,000 emails hacked from the computers of the Democratic National Committee were—shades of Watergate—passed on to WikiLeaks and thus into the public domain.[3]

A Senate inquiry later elaborated on Russia's strategy. It focused on a St. Petersburg enterprise called the "Internet Research Agency." Finding that this agency "sought to influence the 2016 US presidential election by harming Hillary Clinton's chances of success and supporting Donald Trump at the direction of the Kremlin," the Senate's report gave details of how the Internet Research Agency trolled US social media to sow discord in ways that increased Trump's chances of winning the election. Its target of choice was race relations, with 66 percent of its Facebook advertisements focusing on dimensions of that subject. At the time when he discovered the extent of these activities, John Brennan did not know "whether the Russians were simply trying to bloody Hillary Clinton so that she would be weakened politically if she won the November election, as was predicted by most polls, or making a serious effort to get the Republican candidate, Donald Trump, elected."[4]

Debate over Russian intervention intensified in the fall of 2016. Just before the election, the left-wing magazine *Mother Jones* published details from the Steele dossier. The Clinton campaign had hired Christopher Steele, formerly head of the Russian desk at MI6, to do background checks on the Republican candidate. Steele's report detailed the Kremlin's alleged political activities in America and went further. It asserted that the Russians had information about Trump that could be used to blackmail him.

Confronted with these revelations, the Republicans called foul play. In the words of a later report in the *New York Times*, Steele's "research, funded by the Democrats, has been ground zero for conservative allegations that the origins of the Russia investigation were tainted." Trump partisans were incensed that the Steele dossier had been added as an appendix to an Intelligence Community report on Russian meddling.[5] From an early stage, Republicans had pushed back against the idea that the Russians were promoting their candidate. Brennan had to deny that he had prior possession of the Steele dossier and had leaked it to Senate minority leader Harry Reid of Nevada, and stated that he did not see the full text until December. Senate majority leader Mitch McConnell (Republican, Kentucky) was one of those who was suspicious of the role played by Brennan and his CIA. At a briefing on 6 September 2016, McConnell greeted Brennan's remarks about the threat of Russian intervention with the comment, "One might say that the CIA and the Obama administration are making such claims in order to prevent Donald Trump from getting elected president."[6]

A visitor from Mars in the fall of 2016 could have been forgiven for concluding that the CIA and the Republican Party were systemically at loggerheads. The CIA seemed to have entered politics. A thirty-three-year veteran of the CIA who had twice served as its acting director declared for Clinton in the election campaign. Michael Morell added that Trump had endorsed Russian espionage against the United States and was "an unwitting agent of the Russian federation."[7] Never one to shirk a confrontation, Trump issued a string of tweets deriding the CIA as "liberal." He questioned the expertise of the agency and pointed at the 2003 WMD debacle as proof of its incompetence. One journalist asserted, "There is no precedent for such a breach between an incoming president and the nation's national-security professionals."[8]

The bitter debate about Russian intervention continued well into the Trump presidency. Russia would attempt to divert attention from its own "active measures" by alleging that Ukraine was behind the 2016 election trolls. Trump endorsed that charge. He did so in spite of the fact that, by this time, he had had an opportunity to make the CIA his own.[9]

Like Obama before him, Trump chose a new director for the CIA instead of enhancing the agency's apolitical mission by allowing Brennan to stay. This was no surprise as, from Trump's perspective, the outspoken Brennan was poison. The new president's appointee was Mike Pompeo. Not an intelligence professional, Pompeo been first in his class at West Point before studying at Harvard Law School. He had risen to prominence as a Tea Party congressman from Kansas, and there were fears that he would be more loyal to Trump than to the CIA. Pompeo chose Gina Haspel to be his deputy at the CIA. It was a nod in the direction of promoting women, and former CIA director Michael Hayden described her selection as "an inspired choice." Haspel, however, was weighed down by her track record. She was remembered as the person who had ordered the destruction of the CIA torture tapes. She brought with her a ready-made army of critics.[10]

On President Trump's first day in office, as was customary, he crossed the Potomac to visit CIA headquarters. He addressed an assembly of employees in front of the agency's hallowed Memorial Wall, on which were inscribed the names of CIA officers who had died serving their country. His audience must have been nervous. Nobody would have been unaware of his reaction to the Steele dossier: "Intelligence agencies should never have allowed this fake news to 'leak' into the public. One last shot at me. Are we living in Nazi Germany?" Still, the assembled spies were shocked when the new chief executive chose the sacred spot to launch into one of his narcissistic rants—"Probably almost everybody in this room voted for me." They raised a metaphorical eyebrow when Trump declared, "I want to just let you know I am so behind you."

Trump was a master of the short sentence. That, along with his instinctive tonal modulations, made him an effective orator. But he was also a man of notoriously short attention span, and therefore he was potentially a poor customer of an agency like the CIA that was committed to painstaking analysis and used rather longer and more nuanced sentences. He decided to dispense with the President's Daily Brief—without which, Obama had said, a president would be "flying blind." Trump gave the justification, "I don't have to be told—you know, I'm, like, a smart person....I don't have to be told the same thing and the same words every single day for the next eight years."[11]

Trump continued to resist the idea of a written daily briefing. He preferred to be informed by pictures and graphics. His approach fueled the derision of a phalanx of "formers," as intelligence veterans were called. The scornful veterans included former director of national intelligence James Clapper, former directors of the CIA Hayden, Morell (acting), and Brennan, and previous senior CIA officers like John McLaughlin and Paul Pillar. The president's defenders reacted by depicting the critics as hysterical. Trump told his intelligence community to "go back to school." His presidency was a low point in White House–CIA relations.[12]

Trump's Middle Eastern policy was to weaken Iran and to strengthen America's ally Saudi Arabia. Trumpeting a broader dimension of the Trumpian power axis, he encouraged Arab and Muslim nations to forget past enmity and to open diplomatic relations with Israel. Four of them had done so by the end of 2020.

The weakening of Iran involved utilizing the myth of that nation's nuclear program to justify economic sanctions. Just as President Bush had been inconvenienced by the 2007 National Intelligence Estimate indicating that Iran was *not* developing nuclear weapons, so Trump had to disregard estimates that reiterated the finding. Obama had signed up to a "joint comprehensive plan of action" whereby America and its allies would lift economic sanctions imposed on Iran in exchange for Tehran's continued acceptance of non-nuclearization backed up by an inspection regime. The intelligence community affirmed in 2017 that Iran was abiding by the deal, but in May 2018 Trump announced that he would terminate America's compliance with its terms.[13] He was by no means the first president to disregard intelligence findings, but it was still a significant breach with his specialist advisers.

Trump's determination to back the Saudi government knew no bounds. On 2 October 2018 Jamal Khashoggi entered the Saudi consulate in Istanbul to obtain papers relating to his forthcoming marriage. An exile from the Saudi regime who was critical of its effective ruler, Crown Prince Mohammed bin Salman al-Saud, Khashoggi was a columnist for the *Washington Post*. Once he had entered the consulate, Saudi agents overpowered him, killed him, and dismembered his body in anticipation of concealing the remains. By November, the CIA and co-members of the intelligence community had determined that the Crown Prince had ordered the murder. Trump maintained this finding was inconclusive. When Congress voted to suspend America's arms deal with Saudi Arabia, the president vetoed the legislation.

There were developments that could potentially have salvaged the White House–CIA relationship. In March 2018 the CIA was propelled into the limelight when it emerged that it was slated to play a role in improving US–North Korean relations. This was yet another breach of the Hoar Amendment of 1893, which forbade the president from using secret agents in negotiations with foreign powers.[14] But Trump was not engaging in a conspiracy to keep Congress in the dark, the problem that had concerned Hoar over a century earlier. Piling one contradiction on another, Trump openly used a secret agency as a backchannel. His approach was unorthodox.

Kim Jong-un, North Korea's supposedly communist but actually hereditary ruler, was threatening the stability of the surrounding region by developing his country's nuclear weapons arsenal together with the means for their delivery, long-range missiles. In spite of having called Kim the "Little Rocket Man" in one of his knee-jerk remarks, Trump aimed to cultivate a personal relationship with the leader as a prelude to the conclusion of an epochal deal. According to this deal, the United States would lift sanctions on North Korea, and in return that nation would give up its nuclear arsenal and strategic weaponry. When Pompeo agreed to leave the CIA to become secretary of state, he agreed that the CIA could be used as a backchannel to Kim Jong-un. Had the CIA succeeded, no doubt Trump's relationship with the agency would have eased.[15]

Implementation fell to Pompeo's successor at the CIA. On Pompeo's departure from Langley, Gina Haspel became the first female director of the CIA. Her appointment was a symbolic triumph for women, and she was popular within the agency. According to Hayden, "The Agency exhaled when Pompeo was nominated. When Gina was announced, they exulted." Haspel was a thirty-year veteran of the CIA and seemed a safe pair of hands. Unlike Pompeo, she was discreet. Hayden formally endorsed her selection and asserted that "many at Langley believed they had dodged the bullet."[16]

These apparently promising endorsements did not save the day. The overture to North Korea came to nothing. This was not because of any deficiencies on the CIA's part or for that matter Trump's. It was just that the Little Rocket Man loved his toys too much and would not give them up. North Korean diplomacy failed to live up to its promise as a status-saving triumph for the CIA. As for Haspel, her record in regard to enhanced interrogation remained unforgotten. Her tightly sealed lips meant that she continued in post, but they did not enhance her agency's standing. The CIA was destined to endure continuing brickbats from "formers," the media, and Congress, on the one hand, and from Trump, on the other.

In the spring of 2019 Robert Mueller, who had served a twelve-year term as director of the FBI, produced a report commissioned by the attorney general. It confirmed that Russia had meddled in the 2016 election and offered evidence to show that President Trump had obstructed the FBI veteran's inquiry. There was talk of impeachment at that time. As the Republicans controlled the Senate and there was no prospect of conviction, the Democrats held back. Events followed that changed their minds.

On 12 August 2019 an anonymous whistleblower—widely reported in the media as having been a CIA person—leaked details of Trump's 25 July phone call to his Ukrainian counterpart and filed a complaint about it. Trump had asked President Volodymyr Zelensky to find dirt on his leading Democratic rival, Joe Biden, and his son Hunter, who had business commitments in Ukraine. Trump was also accused of threatening to withhold military aid from Ukraine to leverage that country's leaders to do his bidding. The House of Representatives, with its Democratic majority, seized on this episode to impeach Trump in December. The Senate acquitted after the briefest of trials. It was no help to White House–CIA relations that a CIA officer had reportedly played such a key role in the events leading to impeachment.

In December 2019 news began to trickle out of Wuhan, China, about the coronavirus disease that came to be known as Covid-19. When Covid reached America, it caused one of the highest casualty rates in the world. The Trump administration came under fire because the United States had been caught unawares, as well as being unprepared and poorly guided. As in the case of environmental research, it might be questioned whether epidemic prediction is properly the domain of the intelligence community or a matter better left to the scientific and medical communities. Be that as it may, the Department of Defense did contain a National Center for Medical Intelligence, and as early as January 2019 Director of National Intelligence Dan Coats had warned Congress that an epidemic "could lead to massive rates of death." As the Canadian government adviser Wesley Wark later remarked, though, such warnings have little effect unless the agency from which they come can be "persuasive."

Retrospectively, President Trump resorted to a time-honored ruse. He accused the CIA of having been asleep on the job. In particular, he drew attention to his 23 January meeting with the CIA's intelligence briefer Beth Sanner. The briefer had told the president—and CIA officials backed up Trump in his assertion—that the virus was "not a big deal."[17]

Beth Sanner's relationship with Trump is an instructive cameo in itself. In the course of the summer, it emerged that the senior analyst had not briefed Trump about an intelligence finding that the Russians were supplying the Taliban with bounty money to be paid for every US soldier killed in Afghanistan. Sanner explained her omission by saying that she had to be careful to supply her chief customer with intelligence on matters he wanted to hear about. This was not a startling revelation about White House–CIA relations, but it did little to detoxify the existing atmosphere of distrust. Never fully converted to the idea of being briefed, Trump did not ask to see Sanner again after 2 October.[18]

As for Covid, was President Trump justified in his assertion that the CIA and the associated intelligence community should have given him better warning about the catastrophe brewing in Wuhan? Timely warning and preemptive action are essential in the case of rapidly spreading infections. Like the Department of Defense, the CIA had warned—as early as 2008—about the danger of infections spreading from Asia. There were also hints that the CIA gave an early warning in 2019 about China's difficulty in containing the virus.[19]

But Covid was not, except in the fevered imaginings of conspiracy theorists, an act of aggression toward the United States. Perhaps its anticipation was not in the CIA's remit. Arguably, the nation's scientists or the World Health Organization were the people who should have rung the alarm bells more loudly. Yet it could be argued that scientists were not expected to penetrate the secrets of closed societies. The CIA was.

Why was the CIA not subjected to public examination over the issue? Although Covid took more American lives than World War II, the Korean War, and the Vietnam War combined, there was no scrutiny of intelligence performance—as there had been after Pearl Harbor, in reaction to the Vietnam War, and in the aftermath of 9/11. A House Intelligence Committee report released in September did call for a reappraisal of the possibilities of "disease outbreak" in China. It also pointed to the opportunistic use of Covid to justify an expansion of autocratic control in that country. It did not, however, examine the role of the CIA.[20]

Two reasons why the CIA escaped scrutiny spring to mind. First, a real intelligence scandal needs a cover-up, a missing ingredient in the case of the non-anticipation of Covid. Second, the partisan motivation that had triggered scandal in the past was missing. The Republicans were not about to launch a profound attack on the shortcomings of an intelligence system that they themselves had established through the Reform Act of 2004. As for the Democrats, it was not in their interest to distract attention from the culpability of Trump by giving him the excuse of an intelligence failure.[21]

Up to a point, Putin got his way in that his mendacity-mongers helped to invent the Trump presidency. In the closely run presidential contest of 2016, his "troll farms" may have made a difference to the result. The Russian president could rejoice in having installed his man in the White House. Through their media fakery, and because of the bitterness caused by the exposure of that fakery, Putin's agents contributed to the destabilizing polarization of American politics.

President Trump's own obsession with and creation of fake news harmonized with Moscow's efforts. On the other hand, it eroded his domestic political credibility and thus his pursuit of a warmer relationship with Russia. Together with the protracted investigations into Russian electoral meddling, Trump's denials and protestations lowered his believability and destroyed the Putin–Trump dream of improved Russian–American relations. Fake news finally discredited President Trump in the eyes of many voters. They would deliver their verdict in the presidential election of 2020, when the electors' rejection of Trump meant that Putin would lose his point of entry into the White House. In that respect, John Brennan's warning to his Russian counterparts was prophetic. The fake news of the troll farms "backfired."

A victim of presidential disregard and of the fractured nature of American politics, the CIA suffered a dip in its standing in the Trump years. Things could only improve.

18

Back to Work

In February 2021 Thomas P. Bossert was the guest for a "virtual spy chat" organized via Zoom by the International Spy Museum in Washington, D.C. Bossert had been President Trump's original selection as homeland security adviser. In that post, he called for a comprehensive federal program to combat future pandemics. There was little progress on that front, and Bossert quit his post in 2018. When he tried to impress on Trump how serious Covid-19 was after the start of the outbreak, the White House blocked his calls. Speaking in the Spy Museum forum, Tom Bossert conveyed his concern about the state of American preparedness and called the nation's network of nineteen spy agencies a "screwed-up system."[1]

Bossert's view was not unusual, especially amongst CIA diehards who remembered the pre-2004 days when the agency had a higher status. The election of a new president in 2020 in no way guaranteed a return to those good old days. At least there was the prospect, however, that Joe Biden would listen to the CIA's analysts and leave mendacity where it belonged, in the hands of the world's psychological warfare specialists.

To return to work and to protect US security, the CIA needed a change of regime and a return of political stability. One anticipated threat to that security was renewed election interference by Russia. In August 2020 the director of the National Cyber Security Center had issued a warning about expected meddling by China, Iran, and Russia: "We assess that Russia is using a range of measures to primarily denigrate [former] Vice President Biden and what it sees as an anti-Russia 'establishment'."

An intelligence report released in March 2021 confirmed that Russia had tried to discredit Biden and his son Hunter in the course of the

2020 campaign and then had spread false rumors about the election being stolen through ballot rigging. Officers in the CIA and other agencies had collected this evidence in defiance of their superiors, officials appointed by the Trump regime. A separate National Intelligence Council report found that Moscow regarded its meddling as "an equitable response to perceived actions by Washington and an opportunity to both undermine US global standing and influence US decision-making."[2]

Russian meddling did not, however, prove to be a significant factor in the 2020 presidential election. According to one veteran of the intelligence establishment, this was because the National Security Agency ran a successful blocking system. It is also possible that the Russians held back, realizing that their tactics were counterproductive. Moscow may have assumed that the American public were now super-aware of cyber trickery and were unlikely to be fooled again.[3]

Moscow's meddling was at a low ebb, but that did not save the nation from post-election trauma and from fears of a Trumpian putsch. The defeated candidate's refusal to accept the validity of the election result and his claims that there had been voting irregularities culminated in indictments such as that, on 14 January 2021, of Jacob Anthony Chansley, a conspiracy theorist with many aliases:

For the reasons set forth below, the Court should order Chansley to be detained pending trial. Chansley is an active participant in—and has made himself the most prominent symbol of—a violent insurrection that attempted to overthrow the United States Government on January 6, 2021. Chansley has expressed interest in returning to Washington, D.C. for President-elect Biden's inauguration and has the ability to do so if the Court releases him.[4]

Sixth January had been the day set for a joint session of Congress to certify the election of Joe Biden. On that day, the defeated President Trump had addressed a crowd of his supporters:

...our country will be destroyed. And we're not going to stand for that....we are going to try to give our Republicans...the kind of pride and boldness that they need to take back our country. So let's walk down Pennsylvania Avenue.[5]

The crowd did walk down Pennsylvania Avenue and reached the steps of the Capitol building. There, it became a mob that stormed the national legislature. The nation had been dismayed in 1954, when four Puerto Rican nationalists opened fire on the House debating chamber, wounding five congressmen. The shock in 2021 was much greater.

In the course of the violence in which Chansley participated, five people lost their lives for reasons that threatened the stability of the nation.

The Capitol riot led to a second impeachment of President Trump on a charge of "incitement to insurrection," rapidly followed by a second acquittal by the still Republican Senate. The nation was bitterly divided. Sensational rumors circulated. One of them arose from images taken by an alert *Washington Post* photographer. Jabin Botsford was on White House watch duties when his telescopic lens captured the text on notes being carried by a Trump associate. They seemed to suggest that the president was planning to invoke the Insurrection Act of 1807, declaring martial law. Also, the papers suggested that the president was about to move Department of Defense chief of staff Kash Patel, a Trump loyalist, over to the CIA directorship to help oversee a Trumpian coup d'état.[6]

More unsettling than rumors about exotic scenarios was increasing concern about far-right, racist, Trumpian groups like the Proud Boys. Possibly because of lingering Cold War preoccupations with the far left, the FBI had neglected white hate groups. The fact that the Capitol had been caught unprepared and poorly defended was regarded as an intelligence failure, the shortcomings having at least in part come about because the protesters were white and automatically assumed to be patriotic.

Shades of LBJ and Nixon fears about Moscow manipulating the American left, there were suggestions that the Russians might now be stirring up the white racist American right. LBJ and Nixon had instructed the CIA to find a conspiracy that did not exist, but now there seemed to be a better-evidenced case for placing foreign manipulation on the CIA's agenda. Incoming Senate Intelligence Committee chairman Mark Warner (Democrat, Virginia) warned, "We saw Russian entities seed or amplify dangerous narratives that gained significant traction in right-wing media."[7]

Against this background, a very different chief executive entered the White House. There would be no more impatience with intelligence briefings. Infamously, Trump had once insisted on a round of malt shakes to relieve his boredom in mid-briefing session. Campaigners for Joe Biden promised a change in attitude. Tony Blinken, who had been deputy national security adviser under Obama and was now Biden's pick as secretary of state, said, "I know from several conversations with [Biden]...that he has deep concern about what has been done to the [Intelligence Community] these last several years in terms of the politicization, and repairing that starts at the top with the president."[8]

Biden did bring formidable foreign relations experience to his management of the intelligence community. Before becoming Obama's vice president, he had sat in the US Senate, with its constitutionally conferred foreign-policy duties, from 1973 to 2009. Consistently interested in intelligence matters, he had supported Reagan's CIA director Bill Casey by criticizing leakers—and would in 2010 denounce WikiLeaks founder Julian Assange as a "high-tech terrorist"—but had fallen out with Casey over what he depicted as the agency's reckless covert operations.[9] He had twice served terms as chair of the Senate's august Foreign Relations Committee. Biden had left his imprint on a number of policies. For example, in 1992 he had sponsored an amendment forcing President George H. W. Bush to arm the besieged Bosnians.

The incoming president had a long-standing interest in the CIA. While supportive, he had also been critical, especially of Robert Gates. On 5 November 1991 he had voted against Gates's confirmation as CIA director, stating, "I have been disappointed in the past in Mr. Gates's analytical skills, especially in regard to the Soviet Union." On Gates's watch as secretary of defense, Biden, as chairman of the Foreign Relations Committee, demanded the appointment of an independent special counsel to determine the full truth about the destruction of the CIA's torture tapes. Not one to forget such slights, Gates wrote in his 2015 memoir *Duty* that Biden had been "wrong on nearly every major foreign policy and national security issue over the past four decades."[10]

At other times, Biden had supported the CIA and—in a selective way—its operations. In May 2006, amidst the turbulence of intelligence reform, he told Fox TV that he did not want the CIA to be "gobbled up by the Defense Department." As vice president he opposed the bin Laden assassination operation, but he lobbied alongside CIA director Brennan for congressional support for the agency's assistance to the democratic resistance movement in Syria. Biden also had reason to be grateful to the intelligence agency. It became known that when President Trump's attorney Rudolph Giuliani revived the old Ukrainian corruption charges against Biden and his son Hunter in the course of the 2020 election campaign, the CIA warned Trump that the former mayor of New York was being manipulated by Russian agents.[11]

The Biden administration faced a range of intelligence challenges, some of them old, some relatively new, and others neglected by the previous, easily bored incumbent of the White House. Guantánamo, for example, remained a headache inherited from the Bush, Obama, and Trump administrations. Forty prisoners remained in the Cuban facility, transferred there from the proscribed CIA black sites dotted around the globe. Fourteen of the highest-risk languished in a separate prison, where the conditions were atrocious. The cost of keeping Khalid Sheikh Mohammed and the rest in Guantánamo was $13 million per prisoner per year. Congress had forbidden moving the inmates to the United States to be tried in non-military courts and was refusing money to upgrade prison conditions. Khalid Sheikh Mohammed had been sentenced to death for his part in 9/11, but his and others' sentences were caught up in eternal legal wrangling. No other country would take the prisoners, so after more than a decade they were stuck in their hellhole without hope of resolution or release. Biden, like Obama before him, promised to close the facility.[12]

The cyberwarfare threat had become another matter for urgent consideration. The Russians and other nations jealous of or hostile toward the United States were sending wave after wave of attacks. They targeted government, military, and private computer systems. The attacks were displays of strength partly aimed at bullying the United States, showing how the country could be potentially blackmailed. Hostile cybergeeks could implant system bugs to be used for surveillance or to be left as sleepers to be activated later on. Something had to be done. Former homeland security leader Tom Bossert called for the United States to be the world's "sheriff in cyberspace."

The United States remained vulnerable in part because of an old problem—the divide between domestic and foreign intelligence responsibilities. Russians and Chinese were launching their attacks on Fortune 500 firms, utility companies, and government departments by using assets inside the United States. In one notable attack, the Russians reportedly corrupted the systems of the Austin, Texas, company SolarWinds. The Texan company sold products supposedly guaranteeing security to email networks. In December 2020 it was revealed that the email system of the strategically significant Treasury Department had become infected by its purportedly "protective" SolarWinds shield. The FBI and Department of Homeland Security were not equipped to handle the problem. The National Security Agency and the CIA were better equipped, but were prohibited by law from operating internally.[13]

Biden opted to defer a decision on another pressing issue, the use of drones and the American policy of assassinating its enemies. In November 2020 one newsletter commented that the newly elected president had been and remained "silent" on the issue. It will be recalled that drone strikes had become a preferred option under President Obama. Under his aegis there were 550 strikes in Pakistan, Somalia, and Yemen alone. Obama had realized that matters were getting out of hand, with foreign opinion turning against him. He ordered the CIA to publish lists of civilian casualties and then removed from the agency and from the armed forces the license to kill without presidential authorization. Trump removed those restrictions. He also gave local commanders the discretional power to authorize drone strikes outside military zones. Reporters guessed that assassination rates therefore escalated, but the results cannot be determined, as quantification was no longer required.

In early March 2021 President Biden issued "interim guidance" on the use of drone strikes. The administration was making an effort further to quantify the number of civilian casualties caused by drone strikes. Meantime, there were to be no such strikes outside war zones. Defining war zones in an era when there were no more declarations of war and when the enemy was transnational was a potential problem here, but Biden clearly saw a need for restraint in the exercise of power.[14]

Much rested upon Biden's selections to lead the intelligence community. Avril Haines was his pick to be director of national intelligence. Haines had extensive experience of national security governance. She had been the CIA's deputy director, 2013–15, under John Brennan. She had bipartisan support in her confirmation hearings, with senators emitting collective sighs of relief, as Trump had appointed partisan loyalists to the nation's senior intelligence post.

Though Haines was easily confirmed, becoming the first woman in her post, there was potentially trouble on the left. When a legal adviser to the Obama administration, Haines had helped to frame the drone strike program. Her defenders insisted that she had improved transparency and pressed for the achievement of that chimerical goal, lethal explosions with no civilian casualties. But there was a further potential grievance. Haines had supported Trump's appointment of Gina Haspel as CIA director. The left remembered Haspel's role in the destruction of torture tapes.[15]

Biden chose William J. Burns to lead the CIA. Burns was a veteran diplomat who had been US ambassador in Moscow, had under Obama led the secret US delegation to Iran that leveraged Tehran's commitment to nuclear restraint, and had most recently served as president of the Carnegie Endowment for International Peace. He was an outspoken critic of both China and Russia. In a break with the "America First" unilateralist tendency of the previous administration, he advocated security through alliance:

America's partnerships and alliances are what set our country apart from lonelier major powers like China and Russia. For CIA, intelligence partnerships are an increasingly important means of amplifying our understanding and influence.[16]

At his confirmation hearing, Burns endorsed the reorganization of the CIA effected by Brennan. The CIA, he said, should still be dominant in the field of human intelligence (HUMINT), the collection of secret data by undercover agents. It should retain such paramilitary functions as were necessary to the performance of its remit, otherwise leaving them to the military's special units. In his 2019 book *The Back Channel*, Burns had made frequent reference to the CIA, made warm allusions to his CIA friends, and stated that the CIA had a greater impact on Capitol Hill than the Department of State. All these statements were likely to endear him to his future colleagues in Langley.[17]

Burns steered a middle course over the contentious issue of torture. He condemned it but said that as director he would not be pursuing its perpetrators. Committing himself to depoliticize intelligence, he sailed through his confirmation hearing with bipartisan support. Unlike Trump's appointee Mike Pompeo, he would not sit alongside the national security director in the Biden cabinet.[18] Tim Weiner of the *Washington Post* speculated that Biden's appointment marked a departure from the bad old days when:

Political coups…made the CIA's kingmakers a paramount force. It was relatively easy to overthrow Iran, or put the king of Jordan the payroll or run guns to the Afghan guerrillas fighting the Red Army. The man who first set the CIA's clandestine service on the road to secret wars—George Kennan, the most respected diplomat of his era—later called it "the greatest mistake I ever made."...The mission of gathering and analyzing intelligence, and presenting it unvarnished to the president, was and remains the CIA's most vital service. Burns will surely work to renew and strengthen it.[19]

Burns, though he bitterly regretted it, had supported the US invasion of Iraq. There were some doubts as to whether he would now stand up to President Biden if they disagreed about some future crisis. As one journalist put it, "Will Burns buckle under White House political pressure, like, well, George Tenet did when the decision to invade Iraq was in the balance in 2002? Or like DCI Helms, who played his cards so close to his vest on Vietnam that he might as well have had no cards at all, giving the game to the hawks?"[20]

An early test for Burns came when Biden announced he would, by the twentieth anniversary of 9/11, withdraw all US troops from Afghanistan. This put an end to America's longest war and prepared the way for a reorientation of intelligence to concentrate on the intensifying threats to democracy and US interests posed by China, as well as for backdoor diplomacy—with Burns playing a leading role—aimed at achieving security cooperation with Russia. Biden had been a longstanding foe of US military involvement in Afghanistan and had objected when President Obama tried to bring things to an end in that troubled nation by means of a "bulge" in US troop numbers. Burns begged to differ. Fearing that troop withdrawal would weaken the CIA's intelligence-gathering and perhaps permit the terrorists to regroup and strike again, Burns redoubled the CIA's spying efforts, but not until he had declared his opposition to US troop withdrawal.[21]

The announcement that the United States and its allies would withdraw from Afghanistan, which they duly did, sparked a resurgence in Taliban activity just as Burns had feared. As NATO forces withdrew, the theocratic fighters moved into one Afghan city after another, with little resistance. The rapid surrenders reflected a degree of sympathy for the Taliban. The Pew Research Center of Washington, D.C., had indicated following a poll in 2013 that the great majority of Afghans resented the corruption associated with the US-backed regime, saw the Taliban as likely to restore order to their nation, and were disposed to support the Taliban's puritanical sharia regulations. Not for the first time, the problem had arisen that democracy and feminism cannot readily be imposed by the barrel of a gun. So rapid were the collapse of government power and the fall of Kabul by 15 August 2021 that claims and counterclaims circulated about the CIA's record in predicting the surprise. Burns's counsel against US troop withdrawals seemed wise in hindsight, but the intelligence community had not yet confronted with sufficient prescience the significance of the failure of the latest US attempt at nation-building.[22]

As Burns saw it, he had spoken truth to power. It was a quality that his CIA colleagues were likely to respect. It remained to be seen whether he would confirm the pattern. For example, Biden was a long-standing supporter of Israel's leadership and its national security policies. Burns, though, knew at first hand about the plight of the Palestinians. He was mindful of the widespread Middle Eastern support for a two-party state solution to the Israeli–Palestinian dispute. His extensive travel and diplomatic experience had also made him aware of the problem that "The Arab street despises most aspects of American policy, whether in Iraq or Palestine."[23]

There remained, regardless of the quality of Biden's appointees, the question of whether the US intelligence system was systemically "screwed up" post 2004. Certainly there had been some serious predictive failures, among them the non-anticipation of the Arab Spring and, arguably, the Covid menace. But there had been predictive failures before 2004 as well. And since 2004 there had been intelligence successes. The 2007 National Intelligence Estimate on Iran's nuclear capability, delivered in defiance of a main trend in American foreign policy, was one of them. The CIA had also excelled at being an international detective agency, for example, in tracking down bin Laden and in tracing the chain of command in the killing of Jamal Khashoggi that led all the way to the Crown Prince of Saudi Arabia.

The agency, under Presidents George H. W. Bush, Clinton, Obama, and now Biden, was also moving toward greater political correctness in a way that made it more reflective of the American demographic. A hundred days into the new presidency, the CIA ran a social media campaign aimed at recruiting widely and without prejudice. A promotional video showed "Mija," a Latina officer at the agency, walking through the headquarters building in Virginia and declaring, "I am a cisgender millennial, who has been diagnosed with generalized anxiety disorder...I am intersectional, but my existence is not a box-checking exercise." Mercifully, she had recovered from "imposter syndrome" and learned to reject "misguided patriarchal ideas of what a woman can or should be." To conservatives, the clip was "garbage" and evidence of a new liberal takeover at the CIA, while on the left it was charged that the Mija presentation was a "woke" cover-up for sinister policies in the past. In defense of the media ploy, one CIA veteran calmly observed, "Diversity is an operational advantage. Simple as that."[24]

While President Biden's bold domestic initiatives won him the accolade of being the "quiet radical," it remained an open question whether his reforms would restore the standing of the CIA's "quiet Americans." Former director John Brennan observed that the agency would need to be "postured" in a particular way in order to be "valued" by the president. That was an old problem for modern times and carried with it dangers of politicization that had existed before and after 2004 and were unlikely to go away.[25] In July 2021 Biden told his intelligence leaders: "I promise you: You will never see a time, while I'm President, when my administration in any way tries to affect or alter your judgments about what you think the situation we face is. I'll never politicize the work you do."[26] Earlier presidents had made similar promises, just as they had eschewed rash covert operations and then had succumbed to pressure in moments of crisis. Biden gave the impression of being firmer in his resolution than some of his predecessors.

19

Conclusion

The purpose of the CIA in all its seventy-five years of history has been to inform the US president about events, developments, and threats that he and his policymakers might not otherwise be able to perceive. The agency has the further duty of making sense of what it can observe and foresee, which enables it to present cogent interpretations of the data.

It is generally acknowledged that the agency's success has varied according to the abilities of its leaders and staff. That is the premise behind the many barbs it has endured—barbs directed at the real or perceived failings of its personnel.

Slightly less appreciated, if in some respects understood, is the way in which the agency's success has been reliant on its good standing. It is universally agreed that when the White House is deaf to the CIA, the agency is in trouble and is unable to do its job. As with other federal officials, the CIA director's standing in the White House pecking order is a staple of political gossip because it really does matter.

To operate effectively, the CIA must also have the support of Congress, and of public opinion. The latter point is less often explicitly acknowledged. For to appeal to the masses is one thing, but admitting to being an opinion pleaser it quite another. Leon Panetta was exceptional in admitting the role of opinion when he became director of the CIA in 2009 and said his "first goal" was "to restore its standing *with the American public* and political leadership."[1]

In the course of giving a concise history of the agency, this book has unraveled the threads that gave the CIA its mission, as well as those that affected its standing. U-1, as a way of thinking more than as a direct institutional influence, supplied the germinal rationale for the CIA's principal mission, the collection of evidence by secret as well as open means as an aid to analysis. The Research and Analysis Branch of the OSS gave further form to that mission.

The OSS bequeathed the additional legacy of covert operations. This aspect of the CIA's work was secondary to its assigned mission, but was so emphasized in practice that it reduced the standing of the agency's analysts, hurting its prime mission. Covert action's domination of narratives and perceptions of the CIA has reduced the agency's standing as a producer of intelligence. Such activities' impact was international as well as domestic. Over and over again, neutral observers remarked, in the words of a 2014 Senate report, that the agency's covert actions "damaged the United States' standing in the world."[2] Covert actions have affected the global standing of the CIA, the USA, and democracy itself. In our times when "soft power" is so often effective, the reputation of the CIA matters.

The reputation of the CIA is, however, far from uniformly toxic. The leaders of more than a few countries have denounced the CIA in public and availed themselves of its services in private. As one can see from popular culture, the agency is widely held in awe. With regard to the external reach of China and Russia and to how it is exerted, not many citizens of democratic nations would wish the CIA not to exist, and most would wish it to expose malfeasance and to resist it by all reputable means. Democratic oversight of the CIA has made it a model for other nations, has inspired emulation, and has been a spur to the study of secret-intelligence morality.[3]

Reforms have ameliorated the CIA's standing with Congress and the president. In the case of Congress, the reforms of the 1970s produced what one scholar has described as a "grand bargain," though perhaps it resembled more closely a grand power shift. According to the bargain, Congress could prescribe the legal limits for covert operations, oversee the CIA more closely, and require the president to issue formal "findings" to authorize covert action.[4]

The outcome of the reforms was that the CIA's standing with Congress and its new oversight committees became more important. The temper of the times had an effect here. In times of crisis, congressmen and congresswomen rallied to Old Glory and the CIA. But in calmer times, legislators were prone to snipe at the CIA in a manner that eroded its standing, yet could produce reforms that restored that standing.

At all times, of course, legislators on Capitol Hill heed the opinions of those who vote them into office. However, not every intelligence-related controversy stirs the American public. For example, the enduring refrain of those complaining about too much Ivy League influence in the CIA has not struck home. With the exception of Senator McCarthy's denunciations in the early 1950s, there has been no popular rebellion on that issue. It is not, perhaps, a cause for unalloyed congratulation. A "progressive" revolt from below might have stirred the CIA into recruiting, say, Arab Americans, Chinese Americans, and other assets from the lower, ethnic orders in American society, and into making itself an attractive career destination for such groups.

The public has been ambivalent on the use of torture. Even assassinations have failed to stir a popular frenzy. Polls show that the American public disapprove of assassinations in principle, but they have never rated their abolition as a high priority. The one exception was the time when people worried that there was a revenge motive behind the assassination of President Kennedy.

Domestic infractions by the CIA have been a different matter. Such offenses regularly bring out fears of an "American Gestapo" or police state. Dramatic events can also affect public and thus congressional opinion. Covert operations that go spectacularly wrong, such as President Carter's unlucky attempt at hostage rescue, can affect opinion. Other dramas can seal support. In the aftermath of 9/11, and then again following the death of bin Laden, street demonstrators expressed their ecstatic approval by chanting "CIA! CIA!," and their representatives in Congress took note.

In the case of the president, his prior opinion of the CIA, his inclinations, and his sheer ability to absorb, intellectually, what the CIA is trying to tell him all play a role, as do the quality of the CIA's product and the agency's reputation at home and abroad. Personality plays a role. It is a simple question of whether the chief executive gets along with his chief intelligence officer.

The president's opinion of the CIA has been especially important to that institution's standing since the Intelligence Reform Act of 2004, which formally reduced the powers of the agency's director, making the CIA more dependent on presidential favor. To some degree, though, the impact of 2004 needs to be kept in perspective, for there was already a general—and controversial—trend toward short-term analysis, the production of which was still largely entrusted to the CIA after 2004. Gregory Treverton, chair of the National Intelligence Council under President Obama, noted that "the main change" in intelligence analysis "predated the Act, and that is the dominance of the PDB [President's Daily Brief]," by definition an immediate and not forward-looking assessment.[5]

Here, it needs to be said that the history of the analysts is core to the history of the CIA and needs to be winnowed out from it. If this book shows a bias toward the analysts, it is for good reason. Presidents who have turned a receptive ear to the operators while failing to listen to their analysts have done so at their cost, and at a cost to their country.

The same can often be said of presidents who allowed CIA analytical product to be "politicized." Some of the onus for politicization rests with CIA directors who have gone along with twisted findings such as the WMD disgrace. So why has no CIA director ever resigned in protest at the political manipulation of their agency's findings? Mark Lowenthal, a senior veteran of intelligence analysis, notes that three directors resigned because they were disenchanted:

Our history is replete with senior officials speaking out against US policy or simply resigning because they could not support ongoing policy....In 1965, DCI McCone quit as he lost access to President Lyndon Johnson over disagreements about policy in Vietnam. DCIs William Raborn (1965–1966) and James Woolsey (1993–1995) quit because they had no access to the president.[6]

Whereas to resign over loss of standing with the president may be honorable and understandable, however, it is hardly a resignation in protest.

There is, perhaps, one example of a CIA director who *threatened* to resign and got his way. In 1998 the CIA was involved on the security side in US-sponsored Israeli–Palestinian peace negotiations. The negotiations reached their climax at a conference center on the banks of the Wye river, in eastern Maryland. Before agreeing to the deal that President Clinton wanted them to sign, the Israelis threw in the precondition that Jonathan Pollard should be released. US naval intelligence analyst Pollard had received a life sentence in 1986 for spying for Israel. Clinton was inclined to give way to the demand. But George Tenet believed he would lose all credibility with his colleagues at the CIA if there were such a display of leniency to a traitor. He took the president aside in a back room and told him, "If Pollard is released, I will no longer be the Director of Intelligence in the morning." Clinton gave in, the Israelis signed anyway, and Pollard served his prison term until it ended in 2015.[7]

Tenet had confirmed, or so it seemed, his reputation for blunt talk and tough action. Yet he was rolled over, four years later, on a vital issue of analysis—that of Iraq's illusory weapons of mass destruction.

It must be acknowledged that faulty White House policy is a dilemma for CIA directors and indeed for all intelligence leaders. With few exceptions—Bill Casey and Mike Pompeo both sat in presidential cabinets—intelligence leaders are by decree and definition apolitical. Once they speak their minds on a policy issue in public, they are in breach of mission. And there is the danger that if they publicly explain their reasons for dissent, they may well be giving away national security secrets.

An unimpeachable and effective resignation in protest requires considerable skill and sensitivity. Political courage is also an ingredient. Such courage, indeed, can also be admired when it occurs in the lower orders of the agency. Take the case of the CIA analysts who exposed Russian meddling in American politics. They risked both their careers and the standing of their agency by defying their twin bosses, President Trump and the millions who supported him throughout the land. Had it not been for Trump's defeat in the election of 2020, they could have been accused of being in contempt of the democratic principles that had given the CIA its legitimacy in the first place.

Regardless of frequent assertions to the contrary, the CIA has often been politicized. Sometimes, as in Henry Kissinger's data massage in securing SALT I, the politicization has been in a good cause. In other circumstances, it might serve the national interest for a CIA director to be political by resigning to uphold the truth of the day and the reputation of their institution.

Appendix: CIA Directors 1947–2022

(A) THOSE WHO SERVED ALSO AS DIRECTORS OF
CENTRAL INTELLIGENCE

Roscoe H. Hillenkoetter, 18 September 1947 to 7 October 1950
Walter Bedell Smith, 7 October 1950 to 9 February 1953
Allen W. Dulles, 26 February 1953 to 29 November 1961
John A. McCone, 29 November 1961 to 28 April 1965
William F. Raborn, 28 April 1965 to 30 June 1966
Richard M. Helms, 30 June 1966 to 2 February 1973
James R. Schlesinger, 2 February 1973 to 2 July 1973
William E. Colby, 4 September 1973 to 30 January 1976
George H. W. Bush, 30 January 1976 to 20 January 1977
Stansfield Turner, 9 March 1977 to 20 January 1981
William J. Casey, 28 January 1981 to 29 January 1987
William H. Webster, 26 May 1987 to 31 August 1991
Robert M. Gates, 6 November 1991 to 20 January 1993
R. James Woolsey, 5 February 1993 to 10 January 1995
John M. Deutch, 10 May 1995 to 14 December 1996
George J. Tenet, 11 July 1997 to 11 July 2004
Porter J. Goss, 24 September 2004 to 21 April 2005

(B) THOSE WHO SERVED PURELY AS CIA DIRECTORS

Porter J. Goss, 21 April 2005 to 5 May 2006
Michael V. Hayden, 30 May 2006 to 13 February 2009
Leon E. Panetta, 13 February 2009 to 30 June 2011
David Petraeus, 6 September 2013 to 9 November 2012
John Brennan, 8 March 2013 to 20 January 2017
Mike Pompeo, 23 January 2017 to 26 April 2018
Gina Haspel, 21 May 2018 to 20 January 2021
William J. Burns, 19 March 2021–

Abbreviations Used in the Notes

BOHP	George H. W. Bush Oral History Project, MCUV
CCC	Churchill College, Cambridge
CFE	CIA Freedom of Information Act Electronic Reading Room
CJB	Charles Joseph Bonaparte Papers, MLC
CMC	Clark M. Clifford Papers, HST
CREST	CIA Records Research Tool
DNSA	Digital National Security Archive, GWU
FDRL	Franklin D. Roosevelt Library, Hyde Park, New York
FE	Ferdinand Eberstadt Papers, SGM
FRUS	US Department of State, *Foreign Relations of the United States* (Washington, D.C.: Government Printing Office)
GRF	Gerald R. Ford Library, Ann Arbor, MI
GWU	The George Washington University, Washington, D.C.
HDS	Hector Davies Scrapbook, on deposit in SCE
HST	Harry S. Truman Library, Independence, MO
IIS	Institute of International Studies, UCB
JFK	John F. Kennedy Library, Boston, MA
JVF	James V. Forrestal Papers, SGM
K	Kindle pagination
LBJ	Lyndon Baines Johnson Library, Austin, TX
LH	Leland Harrison Papers, MLC
MCUV	Miller Center, University of Virginia
MHL	US Center of Military History Library, Washington, D.C.
MLC	Manuscript Collections, Library of Congress
MMHB	Modern Military Headquarters Branch, NA
NA	National Archives, Washington, D.C.
NA2	National Archives 2, College Park, MD
NRS	National Records of Scotland, Edinburgh
NSA	National Security Archive, The George Washington University
NSFJ	National Security Files, LBJ
OC	Records of the office of the counselor, Department of State, NA
PPF	Post Presidential Files, HST
RAC	Rose A. Conway Files, HST
RDS59	General Records of the Department of State, Record Group 59, NA2

RHJ   Robert H. Jackson Papers, MLC
RHP   Roger Hilsman Papers, JFK
ROHP   Ronald Reagan Oral History Project, MCUV
SCE   Special Collections, Edinburgh University Library, Edinburgh
SCFBI   Small Collections, FBI Files, FDRL
SGM   Seeley G. Mudd Library, Princeton University, Princeton, NJ
UCB   University of California, Berkeley
WCHP   William J. Clinton Presidential History Project, MCUV
WEC   William E. Colby Papers, SGM
WHCF   White House Central Files
WJD   William J. Donovan Papers (microfilm), CCC
WWDA   Woodrow Wilson International Center for Scholars, Smithsonian Institution, Washington, D.C., Digital Archive

Notes

PREFACE

1. The R. Harris Smith papers are located in the Hoover Institution Archives, Stanford, CA.
2. References I retrieved via URLs can be followed up without resorting to error-inducing hieroglyphics. For example, type into a search window the following prose description, or excerpts from it, and you will achieve the desired result: "START 1," ed. James G. Wilson, *FRUS*, 1981–1988, Volume XI (2021) (otherwise, https://history.state.gov/historicaldocuments/frus1981-88v11, accessed on 24 April 2021).
3. The prevalence of socialism in American life is all too easily dismissed as an obsession or opportunist slogan of the American right. See Jeffreys-Jones, *American Left*, 5–7, 10–12, 86–7.

CHAPTER I

1. Wriston, *Executive Agents*, 122–3; Knott, *Secret and Sanctioned*, 50.
2. Both stories in Gilmore's oral recollections in Rawick, ed., *South Carolina Narratives*, 120.
3. Finley quoted in *Congressional Record—House*, 46 Cong. (25 May 1880), 3771–2.
4. Wriston, *Executive Agents*, 296.
5. Letter, Carranza to his cousin Admiral J. G. Ymay, in Wilkie, "Secret Service," 433–6.
6. Wilkie, "Secret Service." See also Wilkie, "Catching Spain's Spies," *Boston Sunday Herald*, 2 October 1898.
7. Dixon, *Clansman*, 103.
8. *Pittsburg Leader*, n.d., in Scrapbooks, vol. 14 (1908), Box 261, CJB.
9. Papen quoted in Jones, *German Spy*, 55.
10. The quotation is from Maugham, *Stories*, vol. 3, 157.
11. Lansing, *War Memoirs*, 318; Stout, "Birth of American Intelligence Culture," 382.
12. Stout, "Birth of American Intelligence Culture," 382–3.
13. Bell to Harrison, 21 January 1918, folder "Miscellaneous Corres. January–August 1918," Box II 115, LH.

14. Harrison communication to Ralph Van Deman, head of the US Military Intelligence Division, 16 November 1917, quoted in Kahn, *Reader of Gentlemen's Mail*, 51.
15. Bell memorandum, 1 May 1919, file "British Intelligence," Box II 102, LH.
16. "Too Much Gratitude?," being ch. 15 in Troy, *Wild Bill and Intrepid: Donovan, Stephenson, and the Origin of the CIA* (New Haven, CT: Yale University Press, 1996): 202–8; riposte in Rhodri Jeffreys-Jones, "Antecedents and Memory as Factors in the Creation of the CIA," *Diplomatic History*, 40/1 (January 2016): 140–54.
17. Stout, "Birth of American Intelligence Culture," 378, 381.
18. See, however, the account of CIA–National Security Agency rivalry in Mainwaring, "Division D."

CHAPTER 2

1. Ray S. Cline, "United States Army in World War II; high command, the Operations Division of the War Department General Staff." 3 vols. (Harvard University Ph.D., 1949).
2. For an example of pluralism within the historian's office at the CIA, see Ludwell Montague's remark about a volume written by his predecessor Arthur Darling, which he described as "a distortion of the history" of the early years of the agency: Montague, *Smith*, xxvii.
3. Cline, *CIA Evolution*, 21, 38, 50.
4. Cline's *CIA Evolution*, published in 1981, reflected a view that was deeply entrenched in official circles, as is evident from Anne Karalekas's history of the CIA, prepared for the mid-1970s Senate inquiry headed by Democratic Senator Frank Church of Idaho. Karalekas drew on seventy-five volumes of internal CIA histories and on sixty interviews with serving and retired CIA officers. She thus presented a widely based perspective when, in explaining the CIA's antecedents, she looked no further than the OSS and the British inspiration behind it: Karalekas, "History," 13 n.1, 16.
5. Ewing, letter to the Secretary of the Admiralty, 27 January 1928, GD433/2/20/2, NRS.
6. *New York Times* editorial, 1 December 1938.
7. Historical branch, G-2, "Materials on the History of Military Intelligence in the United States, 1885–1944" (1944), Part 1, Exhibit B: "Headquarters Personnel and Funds Military Intelligence Activities," MHL.
8. Jeffreys-Jones, *American Espionage*, 125; Angevine, "Gentlemen," 12–13.
9. Departmental order no. 414, 17(?) June 1927, signed by Frank B. Kellogg, and Kirk to Hoover, Director of Naval Intelligence Arthur J. Hepburn and MID assistant chief of staff Stanley S. Ford, 22 June 1927, OC.
10. The quotations, taken from Stimson's diary, are in Kahn, *Reader*, 98.
11. MacPherson, *American Intelligence*, 71; Bruce, *OSS against the Reich*, 207.

12. Adolph Berle, Memorandum of Conversation, "British Relations," 14 September 1939, File 740.00111A/47, Central Decimal File, RDS59, NA; Moynihan statement, Senate Committee on Intelligence hearing, 21 July 1981, 97 Cong., 1 sess., *Intelligence Reform Act of 1981* (Washington, D.C., GPO, 1981), 5–6.

13. Auerbach, *Labor and Liberty*, 99 n.7, 112.

14. Jeffreys-Jones, *Nazi Spy Ring in America*, 77, 87–97.

15. The additional appropriation was for $600,000: Batvinis, *Origins*, 58.

16. Berle, *Rapids*, 320, 404.

17. Hoover memo for Jackson, 5 June 1940, and Jackson memo for Hoover, 22 June 1940, Box 93, folder 5, RHJ.

18. Becker, *FBI in Latin America*, 20–1.

19. Wark, *Ultimate Enemy*, 232–3; Stafford, *Roosevelt and Churchill*, 40–5.

20. Tamm memorandum for director, 3 December 1940, Box I, SCFBI.

21. Clegg and Hince reports, 20 January 1941, and Hoover report to the White House, 5 and 6 March 1941, paraphrased in Charles, "Before the Colonel Arrived," 231–2.

22. Katz, *Foreign Intelligence*, xii–xiii.

23. Watts Hill, memo for Donovan, and Langer, memo for Donovan, both dated 13 November 1944 and in Reel 118, Box 5, WJD. Schlesinger's book *The Age of Jackson* won the 1945 Pulitzer Prize for history.

24. Stafford, *Camp X*, 75–83.

25. Donovan quoted in Waller, *Donovan*, 316.

26. Hector Davies (of SOE), "Allied Mission to the Tarn," eyewitness account, *circa* September 1944, loose-leafed in HDS; Guy de Rouville, Maquis de Vabre, remarks to the author at the Maquis Museum, Vabre, 21 June 2004; Odile de Rouville, letter to the author, 3 December 2011; several emails to the author from Captain LaGueux's widow, Norma, formerly with the CIA, in the year 2012. There are still annual ceremonies in Castres to commemorate the arrival of the OSS in the Tarn.

27. Jeffreys-Jones, *Cloak and Dollar*, 147.

28. Jeffreys-Jones, *Cloak and Dollar*, 149; Colby, "V-E Day in the Norwegian Mountains," typescript in Box 5, WEC.

CHAPTER 3

1. Hadley, *Rising Clamor*, 25.

2. Truman, *Year of Decisions*, 19.

3. Brown, *Last Hero*, 60.

4. Waller, *Wild Bill*, 4.

5. *Chicago Tribune*, 9 February 1945.

6. *Washington Times-Herald*, 12 June 1947.

7. Alvarez, *Spying*, 28.

8. Alvarez, *Spying*, 278–81.

9. Handwritten comment over the initials "HST" on Souers interview transcript, 15 December 1954, PPF Memoirs file (Associates), folder "Souers, Sidney, Dec. 15016, 1954," HST.

10. Draft, "Directive Regarding the Coordination of Intelligence Activities," 1 January 1946, folder "National Security Agency," CMC.

11. Jeffreys-Jones, *CIA and American Democracy*, 35.

12. Souers memo, "Development of Intelligence on the USSR," 29 April 1946, in "Emergence of the Intelligence Establishment," *FRUS*, 344 n. 5.

13. Memo, Vandenberg for president, 24 August 1946, 81–3, and CIG, Office of Research and Evaluation 1/1, "Revised Soviet Tactics in International Affairs," both in Warner, *Cold War Records*, 81–3 and 100–4, n. 5.

14. An enduring Conservative Party myth held that the Army Educational Corps subjected the "Tommies" to "pernicious left-wing propaganda," causing them to vote against Winston Churchill in 1945: Crang, "Politics on Parade," 215.

15. Note on off-the-record press conference, 18 April 1946, general file, folder "Intelligence Service: EAA"; Truman, *Year of Decisions*, 99.

16. James Fenimore Cooper, *The Spy, or, A Tale of Neutral Ground* (New York: Dodd, Mead, 1946 [1821]).

17. See Rowan, *Story of Secret Service*, and Bryan, *Spy in America*.

18. E. R. Warner McCabe lecture transcripts, 1940, in War Department General Staff Military Intelligence Division records, National Archives, Washington, D.C.; "Materials on the History of Military Intelligence in the United States, 1885–1944," U.S. Army Center of Military History library, Forrestal Building, Washington, D.C. In preparation for a forthcoming book, the historian Mark Stout is developing the theme of the military's intelligence history consciousness in the 1920s and 1930s: Mark Stout, "The Enduring Influence of the United States' Intelligence Efforts in World War I," Zoom lecture under the auspices of the North American Society for Intelligence History, 16 June 2020. See also Stout, "World War I."

19. MacPherson, *American Intelligence in War-Time London*, 71.

20. "Remarks by Major General William J. Donovan at Overseas Press Club Dinner," 28 February 1946, Box 31, FE.

21. Enclosure with memo, Donovan to Roosevelt, 18 November 1944, folder "OSS, Donovan – Intelligence Services," Box 15, RAC.

22. Donovan to Forrestal, 14 August 1947, Box 73, JVF.

23. Dulles memo enclosed with Dulles to Senator Chan Gurney (R-SD), chair of the Armed Services Committee, 25 April 1947, in Williams, *Legislative History*, 46.

24. Tydings in the course of hearings before the Senate Committee on Armed Services, 18 March–9 May 1947, in Williams, *Legislative History*, 30; Church

in the course of House debates on the National Security Act of 1947, in Williams, *Legislative History*, 141.

25. McCarthy in the course of hearings before the Senate Committee on Armed Services, 18 March–9 May, 1947, in Williams, *Legislative History*, 43; Robertson in the course of Senate debates on the National Security Act of 1947, in Williams, *Legislative History*, 132, 141; Kendall, "Function," 549.

26. Harry S. Truman, "Truman Deplores Change in CIA Role," *Evansville Courier*, 21 December 1963, and "Limit CIA Role to Intelligence," *Washington Post*, 22 December 1963.

27. Corke, *Covert Operations*, 47, 181 n.12.

28. Timothy Naftali, "Celebrating the Life of Walter Pforzheimer," remarks delivered at Pforzheimer's memorial service, 11 March 2003, in CIA online library; Pforzheimer quoted in Braden, "Birth," 11.

CHAPTER 4

1. Knott, *Secret and Sanctioned*, 73–6 and *passim*.

2. Text of NSC 10/2 in Leary, ed., *Documents*, 131–3; text of NSC 5412/2 in Dylan, *Documents*, 47–51.

3. Gaddis, *Kennan*, 294–5; Truman letter to Democratic Senator Wayne Morse of Oregon, 22 February 1963, PPF Sec. Off. File, "Central Intelligence Agency," HST.

4. On the contributions of socialism to American society, see Jeffreys-Jones, *The American Left*, 6, 11–14.

5. Gleijeses, *Shattered Hope*, 366.

6. Burns, *William Howard Taft and the Philippines*, 41.

7. Gaddis, *Kennan*, 180.

8. Kennan, telegram sent from the American embassy in Moscow to the secretary of state, 22 February 1946, widely published, for example in "Eastern Europe: The Soviet Union," *FRUS*, 696–709.

9. Wilford, *Mighty Wurlitzer*, 226–7.

10. Wilford, *Mighty Wurlitzer*, 154.

11. From the manifesto text in Scott-Smith, *The Politics of Apolitical Culture*, 167.

12. Johnson, *Along This Way*, 327.

13. Thomas Braden, "I'm Glad the CIA is Immoral," *Saturday Evening Post*, 20 May 1967. But Frances Stonor Saunders shows how the CIA ridiculed Picasso's attachment to the left: Saunders, *Who Paid the Piper?*, 68.

14. Memorandum for the record, "Exploitation of Dr. Zhivago," 27 March 1959, Document 19590327, CFE.

15. See, for example, Hugh Wilford's discussion of Frances Stonor Saunders's views: Wilford, *Mighty Wurlitzer*, 102, and cf. Whitney, *Finks*, and Scott-Smith, *The Politics of Apolitical Culture*.

16. Wilford, *CIA, British Left*, 2, 298.
17. Gati, *Failed Illusions*, 73.
18. Burns, *Taft and Philippines*, 8, 12.
19. Mosaddeq quoted in Ghazvinian, *America and Iran*, 144.
20. According to the CIA's Donald Wilber, the principal agency planner of the 1953 coup: Wilber, "Overthrow," Part I, 3.
21. Memorandum by John Waller, Chief of Iran Branch, Near East and Africa Division, Directorate of Plans, CIA, 20 August 1953, "Iran, 1951–1954," *FRUS*.
22. Immerman, *CIA in Guatemala*, 102.
23. CIA memorandum quoted in Kate Doyle and Peter Kornbluh, "CIA and Assassinations: The Guatemala 1954 Documents," National Security Archive Briefing Book No. 4, DNSA.
24. "Program for PBSUCCESS," 12 November 1953, in Cullather, *Secret History*, 152–5.
25. Gerald K. Haines, "CIA and Guatemala Assassination Proposals, 1952–1954" (CIA History Staff Analysis, June 1995), Anonymous, "A Study of Assassination" (n.d.), and "Selection of Individuals for Disposal by Junta Group," 31 March 1954, all in Doyle and Kornbluh, "CIA and Assassinations."
26. "Program for PBSUCCESS," 12 November 1953, in Cullather, *Secret History*, 152–5.
27. *Le Monde*, 2 July 1954, quoted in Gleijeses, *Shattered Hope*, 370.
28. Smith, *Portrait*, 239.
29. "Alleged Assassination Plots," 19, 24–5.
30. For a discussion of the Belgian parliamentary inquiry of 2000–1, see Gerard and Kuklick, *Death in the Congo*, 210–14.
31. For a discussion of this theory see the H-DIPLO Roundtable XXII-4, 21 September 2020.
32. The political scientist Edward Rowe analyzed UN General Assembly voting patterns. He identified a dip in support for the United States in the 1950–5 period (coinciding with the Guatemala and Iran coups), then a recovery in the second half of the decade prior to the years of long-term decline: Rowe, "United States, United Nations," 60.
33. Jameson quoted in Lucas, *Freedom's War*, 108.

CHAPTER 5

1. Some of these failures were more apparent than real and were sometimes the consequences of deaf ears turned to CIA warnings. For a defense of the agency's performances by one of its in-house historians, see McDonald, "CIA and Warning Failures."
2. Kendall, "Function of Intelligence," 549.
3. Kennan, "America and the Russian Future," *Foreign Affairs*, 29 (April 1951), in Kennan, *American Diplomacy*, 124; 1 Corinthians 13:12.

4. Hedley, "Evolution," 23.

5. James M. Doolittle et al., "Report on the Covert Activities of the Central Intelligence Agency," 30 September 1954, in unlabeled box, MMHB, and available in the CIA's digital "reading room."

6. Doel and Needell, "Science," 69–73.

7. Mainwaring, "Division D," 630.

8. Hadley, *Rising Clamor*, 66; Jones, "Journalism," 234.

9. Wilford, *Mighty Wurlitzer*, 120; Jenkins, *CIA in Hollywood*, 1, 15.

10. Willmetts, *In Secrecy's Shadow*, 142–63; McCarthy quoted in Caute, *Great Fear*, 45; figures extrapolated from Barrett, *CIA and Congress*, 156, 219–21.

11. Ronge, *Kriegs- und Industrie-Spionage*, 5.

12. Bernstein, "CIA and Media," 60–3 and 66; Richard M. Bissell, Jr., letter to Lynne Morrison, 13 May 1991, quoted in Morrison, "Journalists," 55.

13. Grose, *Gentleman Spy*, 352; CIA officer quoted in Andrew, *President's Eyes*, 202.

14. Doel and Needell, "Science," 67.

15. For a post-Cold War assessment of whether and how Fuchs might have contributed to the design of Soviet atomic bombs in 1949 and thereafter, see the memoir by the Russian atomic physicist German Goncharov, "American and Soviet H-bomb," 1034–44.

16. Hoover quoted in Sibley, *Red Spies*, 192.

17. "The Berlin Tunnel," 76.

18. "The Berlin Tunnel," 84, 93–4.

19. Murphy, *Battleground Berlin*, 207.

20. Grose, *Gentleman Spy*, 401.

21. Dulles quoted in Stafford, *Spies Beneath Berlin*, 2.

22. Montague, *Smith*, 149–56; Karalekas, "History," 33.

23. "Analyzing Soviet Defense Programs, 1951–1990," anonymized CIA history coded C00872680 and released on 10 September 2014.

24. Laqueur, *World of Secrets*, 147.

CHAPTER 6

1. For one example amongst many of the use of the word "fiasco," see Gleijeses, "Ships in the Night," 1.

2. "Alleged Assassination Plots," 14–15, 24–25, 195; Jones, *Bay of Pigs*, 14; Bohring, *Castro Obsession*, 260.

3. Lorenz, *Spy Who Loved Castro*, vii, 5, 7, 8, 11, 31, 36, 114–21, 228; Jeffreys-Jones, *Nazi Spy Ring in America*, 58; Paul Meskill, story in *New York Daily News*, 13 June 1976; CIA 1977 Task Force Report cited in "Staff Report on the Evolution and Implications of the CIA-Sponsored Assassination Conspiracies against Fidel Castro," for the House Select Committee on Assassinations, 12 March 1979, DNSA, 20; Sam Roberts, obituary, "Marita Lorenz, Who Told Tales of Castro and Kennedy, Dies at 80," *New York Times*, 5 September 2019.

4. CIA 1977 Task Force Report quoted in "Staff Report," 21.
5. Scott, "Incredible Wrongness," 213.
6. "Briefing Papers used by Mr. Dulles and Mr. Bissell—President-Elect Kennedy. COVERT OPERATIONS—CUBA," 18 November 1960, "American Republics," FRUS.
7. Andrew, President's Eyes, 260; Gleijeses, "Ships in the Night," 41; Bohning, Castro Obsession, 52–3; Pfeiffer, "Internal Investigation," 139–46.
8. Mahoney, JFK, 232; Kennedy Press Conferences, 102, 108.
9. The historical quotation is from Immerman, Hidden Hand, 55. The other quotations are from Wyden, Bay of Pigs, 311, and approximately corroborated in Alsop, The Center, 229.
10. Hadley, Rising Clamor, 15, 29; quotations from Alsop, The Center, 214, 228–9.
11. Nixon quoted in Ferris, "Historiography," 111–12; Jeffreys-Jones, "Socio-Educational Composition of the CIA," 421–4; "A New Chief of US Intelligence Takes Over," US News and World Report (11 December 1961), 113; Olmsted, Real Enemies, 125; Powers, Man Who Kept the Secrets, 13–14; Harvey, "Operation MONGOOSE," memorandum for Marshall S. Carter, deputy director of the CIA, 14 August 1962, "American Republics," FRUS.
12. Khrushchev quoted in Nash, Other Missiles of October, 112.
13. Khrushchev quoted in LaFeber, America, Russia, and the Cold War, 221, and in Weisbrot, Maximum Danger, 215.
14. The quotation is borrowed from the title of Duncan Campbell's Unsinkable Aircraft Carrier: American Military Power in Britain.
15. Plokhy, Nuclear Folly, K 214.
16. "Staff Report," 84. On conspiracy theories concerning the JFK assassination, see Olmsted, Real Enemies, ch. 4, 111–48.
17. Johnson, Spy Watching, 261.
18. Mansfield quoted in Washington Post, 17 February 1967; Goldwater quoted in New York Times, 27 February 1967.
19. Katzenbach quoted in Paget, Patriotic Betrayal, 373.
20. Quotation released anonymously by "one of the highest officials in the Kennedy and Johnson administrations," Washington Post, 17 February 1967.
21. Final report of the Katzenbach committee, 29 March 1967, quoted in Saunders, Who Paid the Piper, 405.
22. Wilford, Mighty Wurlitzer, 243; Thomas, Very Best Men, 330, 408 n.33; Nation editorial, 10 April 1967, quoted in Saunders, Who Paid the Piper? 406; transcript, Richard Helms oral history interview, 4 April 1969, by Paige Mulhollan, 41, LBJ.
23. Smith, Declaration, 328; quotation from Smith address in the US Senate: Congressional Record, 23 February 1967, S2498; President Johnson, speaking of J. Edgar Hoover, quoted in New York Times, 31 October 1971.
24. Memorandum, Helms to Rostow, 14 March 1967, folder "Ramparts—NSA—CIA," Subj NSFJ.

25. See Aldrich, "CIA History," 21–6 and *passim*.

26. On the UN, Steven L. B. Jensen (author of *The Making of International Human Rights*), email to the author, 20 October 2020; Bowles, *Promises to Keep*, 392; Morgan, *Anti-Americans*, 9–10. For more extended discussion of the international reaction, see Jeffreys-Jones, *CIA and American Democracy*, 160–3.

CHAPTER 7

1. Allen, *None So Blind*, 235.

2. Immerman, *Hidden Hand*, 76. One CIA historian defined the agency's objectives in Vietnam as bolstering support for the Saigon regime, undermining the Hanoi regime, protecting security, assessing enemy capabilities and intentions, and measuring the progress of the war: Michael Warner, summarized in Dylan, *CIA History, Documents*, 128–9.

3. Conservative revisionists claimed that irresolute leadership in Washington led to the loss of what was a winnable war. See Sharp, *Strategy for Defeat*, 269, 271, and Summers, *On Strategy*.

4. Portes, *Guerre du Vietnam*, 44; Ochiai, "Vietnam Estimate," 46.

5. Thompson, *Make for the Hills*, 125.

6. Memo, "The Situation in South Vietnam," 11 March 1963, folder "Vietnam," RKG Thomson memoranda, 3/63, #1, Box 3, RHP.

7. Ranelagh, *Agency*, 417.

8. Selections in *New York Times*, 31 October 2005, from Robert J. Hanyok, "Spartans in Darkness: American SIGINT and the Indochina War, 1945–1975."

9. McCone, memorandum for the record, "NCS Meeting – 20 Apr 65," 21 April 1965, cited in Ford, *CIA and Vietnam Policymakers*, 79.

10. "Reflection on a Life in Government Service: Conversation with John A. McCone" (Institute of International Studies, University of California Berkeley, synthesis transcription of interviews conducted by Harry Kreisler, 15 October, 3 December 1987, and 21 April 1988), IIS, 5.

11. Memo, Cline to McGeorge Bundy (n.d.), and memo, Cline to McCone "Vietnam" (27 November 1964), both cited in Ochiai, "Vietnam Estimate," 15.

12. "These Men Run the CIA," *Esquire*, 65 (May 1966), 167; author's interview with McGeorge Bundy, 16 May 1984; Cline to the president, 2 March 1966, folder "Central Intelligence Agency," Gen FG 11—2, WHCF, LBJ.

13. Cline, "Factors in Making a Net Assessment of US and Soviet Strategic Forces," memorandum to Undersecretary of State John N. Irwin, 8 March 1971, attached to Cline letter to Kissinger, 24 March 1971, and Kissinger letter to Cline, 31 March 1971, all from LOC-HAK-13-1-31-1, CREST.

14. Final Report, "Intelligence Activities and the Rights of Americans," 100; Memo, Helms to the president, 15 November 1967, enclosing report,

"International Connections of US Peace Groups" (15 November 1967), in folder "US Peace Groups – International Connections," Intelligence File, NSFJ.

15. Lederer and Burdick, *Ugly American*, 273–4; Prados, *Lost Crusader*, 80.

16. Adams, "Vietnam Cover-Up," 62, 63.

17. Colby's biographer John Prados stated that the OSS commander talked the five-man, outnumbered patrol into surrendering. When one of the Germans called out, Colby's men wiped out the entire patrol. Randall Woods, another Colby biographer, wrote that one of the Germans prompted the wipeout when he "raised his pistol." Prados, *Lost Crusader*, 33; Woods, *Shadow Warrior*, 61.

18. Prados, *Lost Crusader*, 33.

19. Quotations from, respectively, Kissinger, *White House Years*, 36; Nixon, *Memoirs*, I: 553; and "Remarks to Top Personnel at the Central Intelligence Agency, March 7, 1969," in Nixon, *Public Papers*, 302.

CHAPTER 8

1. Jeffreys-Jones, *Peace Now*, 54.

2. Prados, *Lost Crusader*, 236–7.

3. See, for example, Colby's story, "V-E Day in the Norwegian Mountains," n.d., *c.*1995, typescript in Box 5, WEC.

4. The foregoing account draws on Kissinger, *White House Years*, 535, and Freedman, *US Intelligence*, 168, as well as on Smith, *Double Talk*, and Prados, *Soviet Estimate*.

5. Fletcher, "Evolution," 319–20, 323, 324. Hunt's fictional character Peter Ward was supposed to be America's answer to James Bond.

6. Story confirmed in posthumously published Sheehan interview with fellow journalist Janny Scott: "How Neil Sheehan Got the Pentagon Papers," *New York Times*, 9 January 2021.

7. Although they would be partially disclosed in the 1970s, the Family Jewels remained mainly secret until 21 June 2007. On that date, the CIA finally released the entire, if still redacted text of the 693-page list also known as the "skeletons": "Family Jewels," memorandum for the executive secretary, CIA management committee, 16 May 1973, in NSA.

8. Helms memorandum on meeting with the president, 15 September 1970, in "Covert Action," 96.

9. Helms review for his top aides, 23 October 2020, quoted in "The CIA and Chile: Anatomy of an Assassination," ed. Peter Kornbluh and Savannah Bock, NSA Briefing Book #728, 22 October 2020.

10. Mondale quoted in Welch, "Secrecy," 170; Harrington to Ford, 17 September 1974, folder "Central Intelligence Agency; 8/9/74–12/13/74," FG 6–2, Subj. WHCF, GRF; Buncher, *CIA and Security Debate*, 11.

11. Marchetti and Marks, *CIA and the Cult of Intelligence*, 42.

12. Marchetti and Marks, *CIA and the Cult of Intelligence*, 40–1; Moran, *Company Confessions*, 113; McGarvey, *CIA Myth and Madness*; Adams, "Vietnam Cover-Up"; Agee, *Inside the Company*; Williams, *Tragedy of American Diplomacy*; Kolko, *Roots of American Foreign Policy*.

13. Angleton left the CIA on Christmas Eve, 1974. See Holzman, *Angleton*, 296.

14. Angleton quoted in Olmsted, *Challenging*, 92.

15. Daniel Schorr discussed the episode in Schorr, *Clearing the Air*, 143–5.

16. *Newsweek*, 16 June 1975, quoted in Johnson, *Season Revisited*, 50.

17. White House, Office of the Deputy Assistant to the President (Richard Cheney), Briefing Book List on Strategy for Dealing with Intelligence-Political Situation, *c.* June 1975, and the excised "Summary of Facts: Investigation of CIA Involvement in Plans to Assassinate Foreign Leaders," 5 June 1975, both from National Security Archive Electronic Briefing Book No. 543, posted 29 February 2016, NSA.

18. Goldwater and Bundy quoted in Johnson, *Season Revisited*, 54, 291; Church quoted in Schorr, *Clearing the Air*, 156.

19. Schorr, *Clearing the Air*, 156. For discussions of maverick CIA behavior, see Olmsted, *Challenging*, 91–6, Johnson, *Spy Watching*, 128–30, and Johnston, *Murder, Inc.*, 152–8. For an assessment of Church's ambitions, see Townley, "Too Responsible to Run."

20. Phillips quoted in Olmsted, *Challenging*, 147.

21. Phillips, *Night Watch*, 237, 278.

22. Moran, *Company Confessions*, 125.

23. *CIA: The Pike Report*, 249–53.

24. Draft, "Speech on Intelligence Decision," 12 February 1976, quoted in Trenta, "Act of Insanity," 137.

25. Walter T. Lloyd on response to questions submitted by the historian M. Todd Bennett. See Bennett's forthcoming book *Glomar: The Ship that Saved CIA Secrecy* (Columbia University Press).

26. Johnson, *Spy Watching*, 141.

27. Barrett, *CIA and Congress*, 459–60.

28. Smist, *Congress Oversees*, 250. See also Olmsted, *Challenging*, 5.

29. Weiner, *Legacy of Ashes*, 349; Jeffreys-Jones, *In Spies We Trust*, 206–7; Christopher quoted in Curran, *Unholy Fury*, 294; John Pilger, "The British-American Coup That Ended Australian Independence," *Guardian*, 23 October 2014.

30. Jeffreys-Jones, *In Spies We Trust*, 169–72; Michel quoted in the Brussels newspaper *Le Soir*, 30 November 2015. On the international impact of the intelligence flap, see Mistry, "Transnational Protest."

CHAPTER 9

1. Gooding, *Rulers and Subjects*; Service, *History*. But at least one historian of Russia has noted the effects of US policy: Westwood, *Endurance*, 492.

2. Reagan, *An American Life*, 548; Gates, *From the Shadows*, 194, 265.

3. See the policy advocated by Defense Secretary Caspar Weinberger, encapsulated in Schweizer, *Victory*, xvii–xviii. There are those who challenge the idea that the United States won the Cold War, arguing that it weakened both antagonists and allowed Germany and Japan to prosper, and that America sacrificed social progress at home because of military spending in its bid to better the USSR. See essays by Alexei Filitov, Geir Lundestad, Gar Alperovitz, and David Reynolds, in Hogan, ed., *End of Cold War*, and Cahn, *Killing Détente*, 196. The quotation is from CIA Directorate of Intelligence, "Analyzing Soviet Defense Programs, 1951–1990," n.d., CIA release 10 September 2014, C00872680.

4. See Philip Taubman, "Cloak and Dagger," a review of Gates, *From the Shadows*, in the *New York Times Book Review*, 19 May 1996, 16, and John M. Broder, " 'Star Wars' First Phase Cost Put at $170 Billion," *Los Angeles Times*, 13 June 1988.

5. Personal note prepared by the deputy secretary of state (Dam), 18 September 1984, "Soviet Union 1983–85," *FRUS*.

6. Kort, *Soviet Colossus*, 291. Martin Anderson, a key member of Reagan's 1980 campaign team and a nuclear-policy adviser, ascribed the arms reductions of the 1980s directly to America's pouring of "hundreds of billions of dollars into increased defense spending": Martin, *Revolution*, 73.

7. See Bearden, *Main Enemy*; Mendez, *Moscow Rules*; and Epstein, *Deception*.

8. Pipes, "Team B," 27; Ellsworth, "Foolish Intelligence," 149.

9. "Views of Principals," attached to "Geneva Arms Control Talks, January 7–8, 1985—Decision Package," National Security Adviser Robert McFarlane, memorandum to President Reagan, n.d., "Soviet Union 1983–5," *FRUS*.

10. Reagan first used the phrase "evil empire" in a speech to the National Association of Evangelicals in Orlando, Florida, on 8 March 1983. He had promised to "unleash" the CIA during his bid for the Republican presidential nomination in 1980: Morton H. Halperin, "The CIA's Distemper," *New Republic* (9 February 1980), 21–2.

11. Barnet, "Balance Sheet," 115; Jeffreys-Jones, *Cloak and Dollar*, 242; US Department of State electronic telegram 324,156 paraphrasing and quoting Radio Moscow, 1 November 1984, in note appended to memorandum of conversation between Secretary of State Schultz and USSR Council of Ministers Chairman Nikolai Tikhonov, 3 November 1984, "Soviet Union 1983–5," *FRUS*.

12. Memorandum from the Vice Chairman of the National Security Council (Herbert E. Meyer) to Director of Central Intelligence Casey, the Deputy Director of Intelligence (John N. McMahon), and the Chairman of the National Security Council (Robert M. Gates), 29 March 1984, "Soviet Union 1983–5," *FRUS*.

13. Quotations from Dujmovic, "Ronald Reagan, Intelligence," 10. Nick Dujmovic was a CIA Russian analyst before becoming its official historian in 2005, and his essay was a contribution to a symposium held under the auspices of the Reagan Presidential Library. Those who remarked on Reagan's enthusiasm for intelligence reports included Secretary of State George Shulz, and national security advisers Richard Allen and Robert McFarlane.

14. Gates, *From the Shadows*, 547.

15. Gaudemans statement, 20 September 1991, "Nomination of Robert M. Gates," 732–40.

16. Ford statement, 20 September 1991, "Nomination of Robert M. Gates," 198–207.

17. Gates, oral history interview with Timothy J. Naftali, 23–24 July 2000, BOHP, 3; Gates, *From the Shadows*, 39; Gates to colleagues in the Departments of Defense and State, and in the Office for National Security Affairs, 11 March 1983, reproduced in "Nomination of Robert M. Gates," 432; "Talking Points Prepared in the Central Intelligence Agency," memorandum attributed to Gates, 19 December 1983, "Soviet Union 1983–5," *FRUS*; Gates, oral history interview with Michael Nelson, 8–9 July 2013, BOHP, 8.

18. Personal note prepared by Deputy Secretary of State Kenneth W. Dam, 18 September 1984, "START 1," *FRUS*.

19. Schweizer, *Victory*, xv–xvii, 35, 42–4. Some Poles must have supported the communist regime of General Wojciech Jaruzelski (1981–9), but the author never met one on his travels in Poland in the course of that decade. In 1989, Solidarity won Poland's first free postwar election.

20. At Langley, there was an acute awareness of the CIA–KGB spy war. See "Talking Points on Spy War Prepared in the Central Intelligence Agency," 19 December 1983, "Soviet Union 1983–5," *FRUS*.

21. CIA, Directorate of Intelligence: "Assessment of Andropov's Power," 13 December 1982, and CIA, Directorate of Intelligence: "Gorbachev, the New Broom," June 1985, reproduced in Dylan, *CIA Documents*, 313–39. See also CIA Intelligence Assessment, "Gorbachev's Economic Agenda: Promises, Potentials, and Pitfalls," September 1985, WWDA document 121,965.

22. "Current Soviet Posture toward the November Meeting," 6 September 1985, and National Security Council minutes, 20 September 1985, both in "Soviet Union 1985–6," *FRUS*.

23. Memorandum, INR Director Morton I. Abramowitz to Shultz, 14 November 1985 (emphasis added), and draft National Intelligence Estimate of 18 November 1985 prepared by "the National Intelligence Officer/USSR and select analysts of CIA/SOVA," both in "Soviet Union 1985–6," *FRUS*.

24. Reagan, *An American Life*, 11–12; Patrick Brogan, "A Little Trouble with Arithmetic," *Observer*, 11 November 1990.

25. Memorandum from Douglas George, the chief of the CIA's arms control staff, to the director of the CIA, 19 August 1986, in "Soviet Union 1985–6," *FRUS*. See also Immerman, *Hidden Hand*, 145.

26. "Guidance for the Upcoming Nuclear and Space Talks Negotiating Round," memorandum from the president's assistant for national security affairs (Frank Carlucci) to President Reagan, 14 January 1987, in "START I," *FRUS*.

27. Webster, oral history interview with Stephen Knott, ROHP, 21 August 2002, 54.

28. Shultz, *Turmoil and Triumph*, 1002.

29. Former CIA officer Bruce D. Berkowitz argues that the CIA acquired its reputation for predictive failure only because, for national security reasons, the analysts' reports remained classified: Berkowitz, "US Estimates of the Soviet Collapse: Reality and Perception," 25.

30. Mendez, *Moscow Rules*, 206–7.

31. Story confirmed by Tony Mendez, formerly the CIA's technical officer in Moscow, as reported in *The Guardian*, 12 October 2002.

32. Webster, oral history interview with Stephen Knott, ROHP, 21 August 2002, 27.

CHAPTER 10

1. Cockburn, *Out of Control*, 214–15; Draper, *Very Thin Line*, 353; Byrne, *Iran-Contra*, 1, 253.

2. *Ash-Shiraa* quoted in Draper, *Very Thin Line*, 457.

3. Bill, *Eagle*, 415. Cf. Elwell-Sutton, *Persian Oil*, 325–6.

4. Donovan, "Intelligence on Iran," 89–95.

5. Ghazvinian, *America and Iran*, 246.

6. Rubin, *Paved*, 45, 109–12, 185; Donovan, "Intelligence on Iran," 92, 127–8.

7. Poll quoted in Rubin, *Paved*, 113.

8. Bill, *Eagle*, 422.

9. See, however, the account given by the Stanford University historian Abbas Milani of Carter's dispatch of a special mission under General Robert Huyser, charged with contacting and assessing Ayatollah Khomeini. As a result of the mission, according to Milani, "The United States began facilitating Khomeini's rise to power" as a man who could bring democracy to his country: Milani, *The Shah*, 394. The University of Pennsylvania's John Ghazvinian gives a different assessment: "Huyser was adamant that Iran's generals would carry the day and preserve the status quo": Ghazvinian, *America and Iran*, 300. Reviewing Ghazvinian's book for the *New York Times* (22 January 2021), Milani wrote, "Evidence declassified in the last few years has shown that in 1979 the United States played a crucial

role in facilitating the clergy's rise to power and, before the hostage crisis of that year, went out of its way to befriend the new regime." Milani emailed the author, 23 January 2021, to explain, without being specific, "Virtually all my evidence about the US role in the ascent of the clergy is based on American archival material." If Milani is correct, the Carter administration was less dogmatic than might otherwise appear to have been the case. One snippet of evidence would appear to support Milani's contention. When on 10 January 1980 Zbigniew Brzezinski sent Carter an overview of US-Iranian relations, 1941–79, the president wrote on top of the memorandum, "Zbig—This is not very helpful to me. I need our decisions recapitulated—i.e., options given to Shah, Huyser's function, etc. J.": "Iran: Hostage Crisis," *FRUS*.

10. CIA national intelligence briefings of 11 May 1978 and unstated day in June 1978, quoted in Donovan, "Intelligence and the Iranian Revolution," 145.

11. CIA National Foreign Intelligence Center, "The Politics of Ayatollah Ruhollah Khomeini," 20 November 1978, slightly redacted version reproduced in Dylan, *CIA Documents*, 201–10. See also Donovan, "Intelligence and the Iranian Revolution," 152, 157.

12. For further appraisal of Brzezinski's stance, see Donovan, "Intelligence and the Iranian Revolution," 149–51.

13. Quotation attributed to Massoumeh Ebtekat in Farber, *Taken Hostage*, 129.

14. Coll, *Ghost Wars*, 42; Zbigniew Brzezinski interview with Vincent Jauvert, published in Jauvert, "Les Révélations d'un ancient conseiller de Carter, 'Oui, la CIA est entrée en Afghanistan avant les Russes'," *Le Nouvel Observateur*, 15 January 1998, 76.

15. Memorandum from Director of Central Intelligence Turner to President Carter, 15 November 1979, in "Iran: Hostage Crisis," document 34, *FRUS*.

16. Memorandum from Director of Central Intelligence Turner to the president's assistant for national security affairs (Brzezinski), 20 November 1979, in "Iran: Hostage Crisis," document 44, *FRUS*.

17. Quotation from memorandum, executive secretary of the department of state (Peter Tarnoff) to Brzezinski, 24 November 1979, and summary of conclusions of a special coordination committee meeting, 1 December 1979, both in "Iran Hostage Crisis," documents 58 and 74 respectively; Farber, *Taken* Hostage, 10. The students' *Documents from the US Espionage Den* appeared in a series of volumes from 1980 onwards.

18. Memorandum prepared by the Iran task force, Central Intelligence Agency, 19 February 1980, in "Iran: Hostage Crisis," document 186, *FRUS*.

19. Casey, remark to congressional staff member and scholar Loch Johnson on 11 June 1984, reproduced in Johnson, *Spy Watching*, 413.

20. Goldwater quoted in *Time* (23 April 1984), 6.

21. Charles A. Briggs, director of the CIA's Office of Legislative Liaison, to Democrat Lee H. Hamilton of Indiana, 28 August 1984, and Casey to

Democrat Anthony C. Beilenson of California, 25 October 1984, both in DNSA; Reagan in United States Information Service transcript of second Reagan–Mondale debate, 21 October 1984.

22. Woodward, *Veil*, 506; Reagan quoted in Lou Cannon, "Reagan Calls Book 'Lot of Fiction' on Casey; President Confirms Secretly Ordering Anti-Terrorist Actions," *Washington Post*, 1 October 1987.

23. Walsh, *Iran-Contra*, 479.

24. Gates, "A CIA Insider Looks at the Battle over Intelligence," *Washington Post*, 29 November 1987.

25. Byrne, *Iran-Contra*, 331.

CHAPTER 11

1. See the numerous courses listed in Marjorie Cline, *Teaching Intelligence in the Mid-1980s*. The United States could be said to have initiated public instruction on intelligence matters.

2. Director of Central Intelligence [William H. Webster], *National Intelligence Daily*, 25 July 1990, heavily redacted text released in 2009 and kindly supplied to the author by Huw Dylan.

3. Quotations from Special National Intelligence Estimate 36.2–90, "Iraq's Saddam Husayn: The Next Six Weeks," 17 December 1990, in Dylan, *CIA Documents*, 347, 358.

4. CIA analysts had predicted the breakup of Yugoslavia, seeing it as inevitable. They favored the recognition of the newly formed states, and they conveyed their finding that Belgrade was secretly supplying Bosnian Serbs with weapons even at a late stage. See National Intelligence Estimate 15–90, "Yugoslavia Transformed," 18 October 1990; "Implications of US Posture on Recognition of Former Yugoslav Republics," 15 January 1992; and DCI Interagency Balkan Task Force Intelligence Force, "Belgrade's Support for the Bosnian Serb Army: Apparently Ongoing," 14 November 1995; all in CFE.

5. Roy, *American Spy*, 31, 105, 152; Richard Aldrich, "Lifting the Lid on a Secret War," *Guardian*, 20 April 2002.

6. Allan Goodman et al., *In from the Cold*, 3, 5; Melvin Goodman, *Failure of Intelligence*, 2; Diamond, *CIA and Culture of Failure*, 15; and Jones, "CIA under Clinton," 504.

7. Vincent Cannistrano, "The CIA Dinosaur," *Washington Post*, 5 September 1991.

8. Herbig, *Espionage by American Citizens*, x.

9. Andrew and Gordievsky, *KGB*, xviii; McIntyre, *Spy and Traitor*, 246; Ben McIntyre, "Escape from Moscow," *Sunday Times*, 16 September 2018. Solzhenitsyn's *Gulag Archipelago*, an exposé of life in Soviet forced labor camps, appeared in Russian in 1973 and in English the following year.

10. The quotation was the consensus view of an Air Force/Defense Intelligence Agency review group in 1979. Polyakov was also betrayed by the FBI traitor Robert Hanssen, and Tolkachev was already in custody when Ames confirmed to his Moscow bosses that he had spied for the CIA. See Bearden, *Main Enemy*, 6, 16, 37, 126.

11. McIntyre, *Spy and Traitor*, 296; Frederick Hitz, review in *Moscow Times*, 14 January 2005, of Victor Cherkashin and Gregory Feifer, *Spy Handler: Memoir of a KGB Officer—The True Story of the Man who Recruited Robert Hanssen and Aldrich Ames* (New York: Basic Books, 2004); Bearden, *Main Enemy*, 157–8.

12. Woolsey quotations from 18 July 1994 address quoted in Wise, *Nightmover*, 317, and R. James Woolsey, oral history interview with Russell L. Riley, WCHP, 13 January 2010, 72–3.

13. Woolsey testimony to the Senate Governmental Affairs Committee quoted in *Washington Post*, 25 February 1993; *New York Times*, 11 March and 2 November 1994.

14. Hodgson, *Gentleman*, 10.

15. Moynihan diary entry, 6 August 1974, quoted in McGurr, "Do We Need the CIA?" 276.

16. Moynihan quoting John le Carré's speech to the Boston Bar Association in 1993, in his remarks promoting S. 126, his Abolition of the Central Intelligence Agency Act, *Congressional Record*, Senate, 104 Cong., 2 sess., vol. 141, S376-377; *New York Times* editorial, "Spies," 1 December 1938.

17. Goodman, "Ending the CIA's Cold War Legacy," 131; Senate Select Committee on Intelligence *Report* 105–24 on appropriations for 1998, 105 Cong., 1 sess., 9 June 1997, 6. The House Permanent Select Committee on Intelligence (*Report* on Intelligence Authorization Act for Fiscal Year 1998, 105 Cong., 1 sess., 18 June 1997, 15) expressed the view that the intelligence community as a whole lacked "the analytic depth, breadth and expertise to monitor political, military, and economic developments worldwide."

18. Roger George, e-interview with the author, 7 May 2021; Woolsey's frequent utterance quoted in Jones, "CIA under Clinton," 409. Roger Z. George is a thirty-year veteran of the CIA who served also on the policymaking staffs of the Departments of State and Defense. He has taught national security at Georgetown University and is a published authority on the theory and practice of intelligence.

19. Some missing pieces of the Venona jigsaw of spy identities became available when in December 1991 the Russian government, now led by Boris Yeltsin, opened the archives of the Communist International and made them available through the new Russian Center for the Preservation and Study of Documents of Recent History. See Haynes and Klehr, *Venona*, 1–2.

20. Holzman, *Angleton*, 319. For a discussion of the voluminous writings on Angleton, with examples of his rehabilitation in the late 1980s and in the

1990s, see Robarge, "Moles," 21–2. A 2006 Hollywood movie, *The Good Shepherd*, starred Matt Damon in an Angleton-like role that re-enacted the half-truths of the Cold War, perhaps reflecting the consensus view that there is no point in being judgmental.

21. Jenkins, *CIA and Hollywood*, 1–16, 19, 73–85.

22. See Fingar, *Reducing Uncertainty*, 2.

23. Doyle, "End of Secrecy," 35; Task Force on Greater CIA Openness, memorandum for the Director of Central Intelligence, 20 December 1991, 1, 2, 15.

24. In 2001 the CIA fired Sterling after he had made frequent complaints about his treatment. In 2003 he told Senate Intelligence Committee staffers that the CIA intended to pass nuclear technology to Iran, and he then explained to journalist James Risen that the plan was to give Tehran flawed blueprints for nuclear warheads. In 2015 he was convicted and imprisoned under the terms of the Espionage Act. Rafia Zakaria, "Why Black Spies Matter," *Al Jazeera*, 24 July 2015.

25. Metzger, "Spies like Us," 12, 15.

26. *New York Times*, 16 May, 6 June 1995.

27. "Some of the Popular Terms of Today's Homosexual Society" (1980), DNSA.

28. *Washington Post*, 5 August 1995; *New York Times*, 9 June 2000. On the Redl case, see Andrew, *Secret World*, 486. Of course, the mythology of the Redl case, incorporating the belief that he was turned on account of his sexuality, may have influenced official opinion. Also of interest is the John Vassall affair. Vassall supplied the Soviets with classified information when he worked in the British navy's Intelligence Division. Arrested in 1962, he served a long prison sentence. Vassall was reputed to have been recruited after succumbing to a gay honeytrap in Moscow. In a 1964 reissue of her classic work *The Meaning of Treason*, the novelist Rebecca West challenged this view. She surmised that Vassall was simply a well-paid professional spy; the homosexuality rationale was an invention designed to indicate that straight men do not commit treason: West, *New Meaning of Treason*, 364–70. Keeping in mind the "lavender scare" of the 1950s, the CIA was no doubt set in its conservative ways by the time of the Vassall case. One might speculate that gay men and women might have an advantage in certain human intelligence settings.

29. Doyle, "End of Secrecy," 35–6, 43, 46.

30. Specter quoted in Johnson, *Threat on the Horizon*, 356; Executive Summary of the commission's report, ch. 14, xxv.

31. "The Growth in Intelligence Funding in the United States, 1980–95, in Constant 1996 Dollars," a graph adapted from the report of the Aspin–Brown Commission, in Johnson, *Threat on the Horizon*, 287; quotation from Tim Weiner report in the *New York Times*, 1 January 1995; Doyle, "End of Secrecy," 46.

CHAPTER 12

1. Bird, *Good Spy*, 173.
2. Quotation from Fred Kaplan, "How the United States Learned to Cyber Sleuth: The Untold Story," *Politico*, 20 March 2016.
3. Tenet, *Center of the Storm*, 100; Coll, *Ghost Wars*, 455.
4. Benjamin, *Age of Sacred Terror*, 248.
5. Clinton interview on Fox News, 22 September 2006, reproduced in Electronic Briefing Book 147, DNSA. President Clinton later recalled that he did not rule out capturing bin Laden alive: "I became intently focused on capturing or killing him and with destroying al Qaeda": Clinton, *My Life*, 798.
6. Dylan, *CIA Documents*, 399; Naftali, "Bush and Terror," 62.
7. Scheuer, *Imperial Hubris*, x; Scheuer, *Osama bin Laden*, ix.
8. Scheuer, *Imperial Hubris*, x, 214.
9. Fanon, *Wretched of the Earth*, 40–1.
10. Tenet, *Center of the Storm*, 492.
11. Scheuer, *Imperial Hubris*, 215–16.
12. Kim Sengupta, review of Scheuer, *Osama bin Laden*, in *Independent on Sunday*, 20 February 2011; quotations from Wilford, *America's Great Game*, xix, xx.
13. Johnson, *Spy Watching*, 16. Loch Johnson served as assistant to Les Aspin, co-chair of the Aspin–Brown Commission, and was a staff member of several intelligence inquiries from the Church investigation onwards.
14. Executive Summary of the commission's report, ch. 7, xxi.
15. Loch Johnson interview with Deutch, 29 October 1998, quoted in Johnson, *Spy Watching*, 421.
16. Director of Central Intelligence Directive 7/3: "Information Operations and Intelligence Community Related Activities" (1 July 1999), archives. gov, 2.
17. United States Commission on National Security, *Road Map for Security*, Appendix 1, "The Recommendations," 118; Johnson, *Spy Watching*, 169, 514 n.80.
18. Johnson, *Threat*, 372.
19. Immerman, *Hidden Hand*, 172; Tenet, *Center of the Storm*, 151, 153.
20. President's Daily Brief, "Bin Laden Determined to Strike in US," 6 August 2001, reproduced in Dylan, *CIA Documents*, 402–4.
21. Robert Baer quoted in Thomas Powers, "The Trouble with the CIA," *New York Review of Books*, 17 January 2002.
22. Coll, *Ghost Wars*, 455–6. 649 n.7.
23. Bob Woodward, review of Tenet's *At the Center of the Storm* in the *Washington Post*, 6 May 2007.

CHAPTER 13

1. Rice, *No Higher Honor*, xiii–xv. For a critical take on Rice's role, see Immerman, *Hidden Hand*, 164, 173.
2. Jeffreys-Jones, *FBI*, 232–3.
3. Scheuer, *Bin Laden*, 17; Scheuer, *Imperial Hubris*, 134. George Tenet, director of the CIA, 1997–2004, agreed that the Israel–Palestine issue was the "root cause of the global terrorism that plagues the world": Tenet, *Center of the Storm*, 54, 60. Tim Naftali, who served as a consultant to the 9/11 Commission, has hinted that Bush had a broader view than Scheuer suggests, writing that the president "viewed al Qaeda more as a symptom than as an illness": Naftali, "Bush and the War on Terror," 62.
4. Panetta, *Worthy Fights*, 270.
5. Naftali, "Bush and the War on Terror," 64.
6. *New York Times*, 27 May 2002.
7. Bush press release of 2002, quoted in Jeffreys-Jones, *CIA*, xxi.
8. Mark Steyn in *The Spectator*, 1 June 2002.
9. *Washington Post*, 19 June 2002.
10. House Permanent Select Committee on Intelligence and the Senate Select Committee on Intelligence, "Report of the Joint Inquiry into the Terrorist Attacks of September 11, 2001," 107 Cong., 2 sess., December 2002, 347–8, 357–9, 363, and supplementary view of Senator DeWine, 12.
11. *9/11 Report*, 358, 411.
12. CIA, "Misreading Intentions: Iraq's Reaction to Inspection Created Picture of Deception" (5 January 2006), 1, reproduced in Tom Blanton, "A Classified Mea Culpa on Iraq," *FP/Foreign Policy*, 5 September 2012.
13. Quotation from Bamford, *Pretext for War*, 334.
14. Tenet, *Center of the Storm*, 480; Immerman, *Hidden Hand*, 179, 180; National Intelligence Estimate, "Iraq's Continuing Programs for Weapons of Mass Destruction," October 2002, in Prados, *Hoodwinked*, 51–76.
15. Rice, *No Higher Honor*, 197–8.
16. Faddis, *CIA War in Kurdistan*, 215.
17. Joseph Wilson, "What I Didn't Find in Africa," *New York Times*, 6 July 2003.
18. Tenet, *Center of the Storm*, xv, 150–3, 479, 480.
19. Tenet, *Center of the Storm*, xv.
20. Select Committee on Intelligence, United States Senate, *Report on the U.S. Intelligence Community's Prewar Intelligence Assessments on Iraq*, 108 Cong., 7 July 2004, 14, 24.
21. *New York Times*, 15 April 2004, reporting an interim 9/11 commission report.
22. *9/11 Report*, 408–15.
23. The words are those of Richard Immerman, who served in the Bush administration, 2007–8, as Assistant Deputy Director of National Intelligence for Analytic Integrity and Standards, and later wrote a history

that embraced the period and is authoritative in spite of redactions imposed by the CIA's prepublication reviewers: Immerman, *Hidden Hand*, 189.

24. Ransom, "Secret Intelligence," 225. Brent Durbin offers a different view, that intelligence reform needs political consensus at home: Durbin, *CIA and the Politics of U.S. Intelligence Reform*, 13–14.

25. Bush, *Decision Points*, 4, 84.

26. Undated fax, George H. W. Bush to George W. Bush, quoted in Bush, *Decision Points*, 225.

27. Bush, *Decision Points*, 19.

28. According to Gregory Treverton, a leading intelligence consultant and authority who directed the National Intelligence Authority under President Obama, "the main chain [of change] predated the [2004] Act, and that is the dominance of the PDB [Presidents' Daily Brief]":Treverton e-interview with the author, 16 April 2021.

29. Quotations from Johnson, *Threat on the Horizon*, 391, 478 nn.28, 29.

30. Author's conversation with Art Hulnick, Boston, 21 October 2003. Arthur S. Hulnick was a twenty-eight-year veteran of the CIA who kept in touch with his former colleagues when he became an intelligence and international relations professor at Boston University.

31. Hulnick, *Keeping Us Safe*, 189.

32. Posner, *Preventing Surprise Attacks*, 4, 67–8; Laurence H. Silberman et al., letter to the president transmitting the *Report of the Commission on the Intelligence Capabilities of the United States Regarding Weapons of Mass Destruction*, and summarizing its findings, 31 March 2005.

33. Johnson, *Threat on the Horizon*, 403.

34. *The Economist*, 30 May 2020.

CHAPTER 14

1. Whipple, *Spymaster*, 216.

2. Author's e-interview with Fingar, 13 February 2021.

3. Fingar, *Reducing Uncertainty*, 138.

4. Fingar, *Reducing Uncertainty*, 7.

5. Fingar, *From Mandate to Blueprint*, 122–9.

6. Kendall, "Function," 549; Fingar, *Reducing Uncertainty*, 90; quotation from Roger Z. George, e-interview with the author, 7 May 2021; Christopher Kojm, e-interview with the author, 27 February 2021; Gregory Treverton, e-interview with the author, 16 April 2021. Kojm served on the staff of the House Foreign Affairs Committee, as deputy assistant secretary of state in INR, and as deputy director of the 9/11 Commission before chairing the National Intelligence Council, 2009–14.

7. Keefe, "Privatized Spying," 303; James Bamford, "This Spy for Rent," *New York Times*, 14 June 2004; Priest, *Top Secret America*, 192.

8. Keefe, "Privatized Spying," 297; Jenkins, *CIA in Hollywood*, 20.

9. Puyvelde, "Sécurité," 25; Fingar, *Reducing Uncertainty*, 10, 51.

10. Quotations from Fingar, *Reducing Uncertainty*, 58, 60; remarks on pandemic forecasts in *Global Trends 2008* by Roger George in George e-interview with the author, 7 May 2021.

11. Rumsfeld quoted in Ghazvinian, *America and Iran*, 459.

12. Arnold, "Iran Nuclear Archive," 230.

13. Ghazvinian, *America and Iran*, 469.

14. Hoekstra quoted in *Washington Times*, 19 January 2010; other quotations from Fingar, *Reducing Uncertainty*, 123; Immerman, "Tradecraft," 217.

15. Rovner, *Fixing the Facts*, 34.

16. Quotations from Ghazvinian, *America and Iran*, 470; Rovner, "Preparing"; William J. Broad and David E. Sanger, "In Nuclear Net's Undoing, a Web of Shadowy Deals," *New York Times*, 25 August 2008.

17. Maureen Dowd, op-ed, "Seven Days in December?" *New York Times*, 5 December 2007.

18. Immerman, *Hidden Hand*, 208.

19. Whipple, *Spymaster*, 216–7, 223.

20. Hayden, *Playing to the Edge*, 243, 245, 248; Panetta, *Worthy Fights*, 2, 7, 10–11, 221.

21. *Report of the Commission to Assess the Ballistic Missile Threat to the United States*, 104 Cong., 1998, II, Executive summary. A. Conclusions of the Commissioners, paragraph 1.

22. Comment in *The Times*, 23 March 2019, on Rumsfeld statement in press briefing of 12 February 2002.

23. Stephan A. Cambone memorandum for the secretary of defense, 30 September 2004, quoted and paraphrased in Mazzetti, *Way of the Knife*, 81; Former assistant secretary of defense (1981–5) Lawrence J. Korb and Jonathan D. Tepperman, "Soldiers Should Not Be Spying," *New York Times*, 21 August 2002.

24. Walter Pincus, "Pentagon May Get New Intelligence Chief," *Washington Post*, 19 August 2002.

25. Mazzetti, *Way of the Knife*, 75–6; quotation from Clapper, *Facts and Fears*, 107; Thom Shanker and Mark Mazzetti, "New Defense Chief Eases Relations Rumsfeld Bruised," *New York Times*, 12 March 2007.

26. Gregory Treverton (chairman, National Intelligence Council, 2014–17), e-interview with the author, 16 April 2021. Cf. Christopher Kojm (chairman, National Intelligence Council, 2009–14), e-interview with the author, 27 February 2021: the CIA's "drafting contribution to NIC products significantly exceeds the contribution of any other IC [intelligence community] agency."

CHAPTER 15

1. Gregory Treverton, e-interview with the author, 16 April 2021. The scholar Adam Jacobsen argues that the problem lay not with policy but with poor implementation due to a lack of controls: Jacobson, "Back to the Dark Side," 226.
2. Risen, *State of War*, 24; Executive Order 13440: Interpretation of the Geneva Conventions Common Article 3 as Applied to a Program of Detention and Interrogation by the Central Intelligence Agency (20 July 2007), Section 3, (i) (A) and (B).
3. Best, "Special Operations Forces," 1, 3.
4. *Washington Times*, 4 February 2003; *Chicago Tribune*, 7 January 2007.
5. Nathan Hodge, "CIA's Predatory [*sic*] Behavior is Cause for Concern," *Newsday*, 7 June 2002.
6. Ball op-ed in the *New York Times* of 16 December 1984, quoted in Fuller, *Shoot It*, 33; Gregory Treverton e-interview with the author, 16 April 2021.
7. Schmitt, "State-Sponsored Assassination," 683, 685; italicization in the original in Newman and Van Geel, "Executive Order 12,333," 447. The issue was also discussed in the *Hamline Journal of Public Law and Policy* (Summer 1992), *Temple International and Comparative Law Journal* (Fall 1991), and seven other law journals to the author's knowledge.
8. Quotations from Fuller, *Shoot It*, 144–5.
9. George Brant's 2012 *Grounded*, viewed by the author at the Traverse Theatre, Edinburgh. For this play Brant received, amongst other accolades, the Fringe First Award at the Edinburgh Fringe Festival.
10. A close academic study of public opinion in the Bush years indicated that, in spite of a widespread belief to the contrary, the American people disapproved of the use of torture, even when in the interest of national security. In a perverse twist, opinion would switch to supporting torture after President Obama banned its use: Gronke, "US Public Opinion on Torture, 2001–2009," 442.
11. In making this observation, the historian Christopher Andrew noted President Bush's denial that waterboarding was torture and observed that the same method of eliciting information had been used by American forces during the American occupation of the Philippines during the presidency of Theodore Roosevelt. Andrew, *Secret World*, 116–17.
12. Email to the author from an anonymous British Army officer, 23 April 2021.
13. *New York Times*, 22 January 2020; *Christian Science Monitor*, 27 June 2004; Morell, *Great War of Our Time*, 271.
14. Scott Shane and Mark Mazzetti, "Tapes by CIA Lived and Died to Save Image," *New York Times*, 30 December 2007.

15. Quotations from redacted secret cable, "Request Approval for [Excised] Video Tapes," 8 November 2005; destruction date given in redacted top secret cable, "Destruction of [Excised] Video Tapes," 9 November 2005, both cables in DNSA.
16. Morell, *Great War of Our Time*, 259–60; quotation from Whipple, *Spymaster*, 221.
17. Whipple, *Spymaster*, 197, 302.
18. Deborah Solomon, "Questions for Melissa Boyle Mahle, Agent Provocateur," *New York Times Magazine*, 17 April 2005. On 22 January 2003 the CIA advertised for Arab-American recruits: *New York Times*, 23 January 2003. On women making a difference to foreign policy: Jeffreys-Jones, *Changing Differences*.
19. Tenet, *Center of the Storm*, 54; Aluf Benn, "CIA Sets Up Department to Implement the Road Map," *Haaretz*, 24 March 2003.
20. Serafino, *US Occupation Assistance*, 1.
21. Naftali, "Bush and War on Terror," 85. Naftali wrote the 9/11 Commission's report on counterterrorism, later developed into the book *Blind Spot*.
22. Both quotations from Risen, *State of War*, 22–4.
23. Remarks by Secretary of State Rice en route to Berlin, Germany, released 5 December 2005, DNSA.
24. Cumming, *Covert Action*, 1.
25. Rice, letter to Republican Senator John Warner of Virginia, 16 February 2007, DNSA.
26. "Extraordinary Rendition," 1, 4, 8, 13.
27. *New York Times* editorial, 2 March 2008; Mark Mansfield, director of public affairs, CIA, to *New York Times*, letter published on 9 March 2008.
28. Senator Feingold introduced her Executive Order Integrity Act on 31 July 2008: *Congressional Record*, Senate, S7955-6.
29. "Committee Study," xv.
30. William Glaberson, "6 at Guantánamo Said to face Trail in 9/11 Case," *New York Times*, 9 February 2008.
31. Scott Shane, "CIA Chief Doubts Tactic to Interrogate is Still Legal," *New York Times*, 8 February 2008.
32. Mark Mazzetti and Scott Shane, "After Sharp Words on CIA, Obama Faces Delicate Task," *New York Times*, 3 December 2008.
33. Pew Research Center, "Top Issues for 2008," 21 August 2008; Mason and Morgan, *Seeking a New Majority*, 220.

CHAPTER 16

1. Ransom, "Secret Intelligence," 220. Emphasis in the original.
2. Text of Reducing Over-Classification Act, H.R. 553, *Congressional Record*, 3 February 2009, House, H893–98 (bipartisan law finally signed by President Obama on 7 October 2010); Obama, *Promised Land*, 314, 354;

Obama speech in swearing-in ceremony recorded in *Secrecy News*, 2009/7, 22 January 2009.

3. Panetta, *Worthy Fights*, 2–10, 221.

4. Charlie Allen quoted in Whipple, *Spymasters*, 243; Panetta, *Worthy Fights*, 270; Obama, *Promised Land*, 312.

5. Christopher Kojm e-interview with the author, 27 February 2021; George, "Reflections on CIA Analysis: Is it Finished?" 79; Gregory Treverton e-interview with the author, 16 April 2021.

6. Treverton, "Conclusion," in Hutchings and Treverton, *Truth to Power*, 200. In the same volume, Hutchings traced the National Intelligence Council's precursor histories back through the "golden age" of US intelligence in the 1950s: Hutchings, "Introduction," 7. The present-day title of the National Intelligence Council dates from 1979, and the 2004 Reform Act revitalized its mission. Robert Hutchings led the organization from 2003 to 2005, and Gregory Treverton from 2014 to 2017.

7. Nolan, "Information sharing," 158–9.

8. Federation of American Scientists Intelligence Resource Program, "Intelligence Budget Data [2005–2020]," n.d.; Erwin, *Intelligence Spending*, 2.

9. *Washington Post*, 29 August 2013. The document was one of the Edward Snowden disclosures.

10. Quotation from the pre-delivery text of John Brennan, "The Ethics and Efficacy of the President's Counterterrorism Strategy," address to be delivered at the Woodrow Wilson International Center for Scholars, Washington, D.C., 30 April 2012, DNSA.

11. Fuller, *See It*, 215.

12. Fuller, *See It*, 214; quotations from Immerman, *Hidden Hand*, 223, 224.

13. Both quotations from Conor Friedersdorf, "Obama's Weak Defense of his Record on Drone Killings," *The Atlantic*, 23 December 2016.

14. Whipple, *Spymasters*, 246–9.

15. Department of Justice White Paper, draft, "Lawfulness of a Lethal Operation Directed against a US Citizen Who Is a Senior Operational Leader of Al-Qa'ida or an Associated Force," 8 November 2011, 16.

16. Maureen Dowd, "The CIA's Angry Birds," *New York Times*, 16 April 2013.

17. Scheuer quoted in Fuller, *See It*, 246.

18. Figures cited in Fuller, *See It*, 230.

19. George, "Reflections on CIA Analysis," 77.

20. Immerman, *Hidden Hand*, 224; Hutchings, "Intelligence Integration," 169; quotation from Jeremy Bowen, "How the Dream Died," *New Statesman*, 26 February–4 March 2021, 21.

21. *New York Times*, 25 August 2008.

22. "Review of the Terrorist Attacks," 16.

23. *New York Times*, 21 June 2012, 14 October 2014.

24. Panetta, *Worthy Fights*, 289; Obama, *Promised Land*, 676–7.

25. Morell, *Great War*, 272. An alternative account of how bin Laden was tracked down stresses non-coercive interrogations by the FBI and the correlation of information from thousands of al-Qaeda documents captured in battle: Bergen, *Manhunt*, 93.

26. Via conference call, White House Press Briefing by Senior Administration Officials on the Killing of Osama bin Laden, 2 May 2011.

27. Obama, *Promised Land*, 681–2; Immerman, *Hidden Hand*, 217.

28. US intelligence officer who has to remain anonymous, email to the author, 7 May 2020. According to the author's late father, T. I. Jeffreys-Jones, he slept with a pistol in his hand when on dangerous operations on behalf of the British Army in North Africa in World War II.

29. "Background Briefing with Senior Intelligence Official at the Pentagon on Intelligence Aspects of the U.S. Operation Involving Osama Bin Laden," 7 May 2011, DNSA.

30. Rollins, *Osama bin Laden's Death*, 5–6; quotation from Whipple, *Spymasters*, 254.

31. Author's e-interview with Kojm, 26 February 2021.

32. *New York Times*, 9, 10, 13 November 2012.

33. Gregory D. Jensen, "The Untouchable John Brennan," *BuzzFeed News*, 24 April 2015.

34. Brennan, "Human Rights," 89, 104, 110. Brennan's thesis adviser in UT Austin was Carl Leiden, an authority on Middle Eastern politics who had an academic interest in assassination. Sadat was assassinated by a self-declared Islamist on 6 October 1981. Hosni Mubarak succeeded him and governed Egypt until removed in the Arab Spring uprising, on 11 February 2011.

35. Greg Myre, "After Chasing Threats Abroad, Former CIA Chief John Brennan Says the Risk Is at Home," National Public Radio, 5 October 2020; quotation from Fuller, *See It*, 212.

36. "Study Overseen by Vernon Jordan '57 Finds CIA's Diversity Efforts Lacking," DePauw University press release, 30 June 2015.

37. Mark Mazzetti, "CIA to Focus More on Spying, a Difficult Shift," *New York Times*, 23 May 2013.

38. "Information Technology," 12, 77.

39. All quotations from Scott Shane, "CIA's History Poses Hurdles for a Nominee," *New York Times*, 6 March 2013.

40. *New York Times*, 19 July 2013.

41. *New York Times*, 20 March 2014.

42. "Committee Study," xi, xii, xiv, xx, xxv.

43. All quotations from Miles, *Perspectives*, 7, 11, 13 n.78.

44. Mark Mazzetti, "CIA Study of Covert Aid Fueled Skepticism about Helping Syrian Rebels," *New York Times*, 14 October 2014.

45. Obama, *Promised Land*, 699.

CHAPTER 17

1. Brennan, *Undaunted*, K372. According to one estimate, the United States "meddled in the affairs of foreign countries on 62 different occasions during the Cold War": Walton, "Spies, Election Meddling," 120.

2. "Net Favorability of the CIA, by Party Affiliation [2003–17]," NBC News/Wall Street Journal poll published on 4 January 2017. A Gallup poll of 2 January 2017 showed that the CIA was the sixth most popular federal agency out of a list of thirteen. A year later, the Pew Research Center had the CIA fourth on a list of ten federal agencies, with the Democrats still more favorable than the Republicans, but by only a four-point margin.

3. Walton, "Spies, Election Meddling," 107–23; Julian E. Barnes, "CIA Informant Extracted from Russia Had Sent Secrets to US for Decades," *New York Times*, 9 September 2019; quotation from Brennan, *Undaunted*, K364.

4. "Russian Active Measures," 4, 6; Brennan, *Undaunted*, K364.

5. Katie Banner, "[US investigating attorney John H.] Durham is Scrutinizing Ex-CIA Director's Role in Russian Interference Findings," *New York Times*, 19 December 2019. President Trump's appointee, Attorney General William Barr, assigned Durham to investigate the FBI's investigation of alleged Russian interference in the 2016 election.

6. McConnell quoted in Brennan, *Undaunted*, K365.

7. Morell, "I Ran the CIA. Now I'm Endorsing Hillary Clinton," *New York Times*, 5 August 2016.

8. David Leonhardt, "The CIA, Just Another Liberal Enemy," *New York Times*, 19 December 2016.

9. *New York Times*, 22 November 2019.

10. Quotation from Hayden, *Assault on Intelligence*, 94.

11. All quotations from Dylan, *CIA*, 493–4.

12. Quotation from Barrett, "Retired Intelligence Leaders versus Donald Trump," 4. See also Immerman, "Trump to the Intelligence Community".

13. Evidence would appear to point to another of Trump's anti-Iranian actions, the assassination by drone strike of Iran's military chief Qasem Soleimani on 3 January 2020, an action possibly aimed at frustrating any effort by the incoming Biden administration to improve relations with Iran. For an assessment of how the CIA was involved in the planning and execution of the deed, see Jack Murphy and Zach Dorfman, " 'Conspiracy Is Hard': Inside the Trump Administration's Secret Plan to Kill Qassem Soleimani," *Yahoo! News*, 8 May 2021.

14. See Chapter 1, and Wriston, *Executive Agents*, 296.

15. Mark Landler, "Spies, Not Diplomats, Take Lead Role on Planning Trump's North Korea Meeting," *New York Times*, 16 March 2018.

16. Hayden, *Assault on Intelligence*, 94.

17. Coats quoted in Erik H. Dahl, "Was the Coronavirus Outbreak an Intelligence Failure?" *The Conversation*, 16 June 2020; Wark remark in the course of his Zoom presentation, "The Canadian Intelligence System and Global Pandemics," North American Society for Intelligence History brown bag seminar, 21 September 2021; Trump quotation paraphrasing Sanner and other intelligence community officials in *New York Times*, 21 May 2020.

18. Natasha Bertrand, "Trump's Intel Briefer Breaks Her Silence," *Politico*, 6 June 2020; *New York Times*, 30 October 2020.

19. "It's my understanding that CIA provided early warning of the outbreak not being contained in China and anticipating it would spread": Roger George, e-interview with the author, 7 May 2021. George retired from the CIA in 2009, but remained in touch with some of its officials.

20. "China Deep Dive," 6. There was, however, discussion of the problem in a special issue of the journal *Intelligence and National Security*, 35/4 (2020).

21. The Ransom thesis, that criticism of the CIA intensifies in times of relative international harmony, seems irrelevant to the Covid case.

CHAPTER 18

1. Harry Cockburn, "Coronavirus: White House 'blocking calls to Trump' from former adviser trying to warn him how dire pandemic is," *Independent*, 13 March 2020; Bossert remarks, 25 February 2021.

2. Julian E. Barnes, "Russian Interference in 2020 Included Influencing Trump Associates, Report Says," *New York Times*, 16 March 2021; National Intelligence Council, "Foreign Threats," 5.

3. William Evanina, "Election Threat Update for the American Public," news release, 7 August 2020 (according to Evanina, China favored Trump, but Iran opposed him); Chris Costa (recently counterterrorism director for the National Security Council), remarks during a "virtual spy chat" organized via Zoom by the International Spy Museum, of which he was director, 14 January 2021.

4. Government's Brief in Support of Detention, United States District Court for the District of Arizona, 14 January 2021.

5. Trump quoted in the *Washington Post*, 7 January 2021.

6. Maggie Haberman, "Photos Capture Notes from Trump Ally Leaving the White House on Friday," *New York Times*, 15 January 2021.

7. Mark R. Warner, "Statement of Incoming Senate Intel Chairman on New Sanctions against Russia-Linked Disinformation Network," press release, 11 January 2021.

8. Blinken quoted in Natasha Bertrand, "Biden Would Revamp Fraying Intel Community," *Politico*, 19 October 2020.

9. David Usborne, "Assange Is a 'Hi-Tech Terrorist', says Biden," *Independent*, 20 December 2010.

10. Biden quoted in *New York Times*, 18 January 2014; *New York Times*, 9 December 2007; Gates, *Duty*, 288.

11. Biden quoted in *New York Times*, 7 May 2006; *New York Times*, 15 July 2013, 16 October 2020, and 16 March 2021.

12. *New York Times*, 15 December 2020 and 15 February 2021.

13. Bossert remarks, 25 February 2021; *New York Times*, 14 March 2021.

14. Elise Swain, "Joe Biden's Silence on Ending the Drone Wars," *The Intercept*, 22 November 2020; editorial, BloombergQuint, 16 March 2021; Charlie Savage and Eric Schmitt, "Biden Secretly Limits Counterterrorism Drone Strikes away from War Zones," *New York Times*, 3 March 2021; Charlie Savage, "Trump's Secret Rules for Drone Strikes outside War Zones Are Disclosed," *New York Times*, 6 May 2021.

15. *New York Times*, 23 November 2020 and 21 January 2021.

16. Quotation from Statement for the Record [opening statement at confirmation hearings], Senate Select Committee on Intelligence, Director of CIA Nominee William J. Burns, February 2, 2021.

17. Burns, *Back Channel*, K423. There were twenty-two references to the CIA in the text of this autobiographical survey of Burns's diplomatic career.

18. *Guardian*, 12 January 2021; Zachary Cohen, "Biden's CIA Director Skates through Confirmation Hearing and Receives Bipartisan Praise," CNN, 24 January 2021.

19. Tim Weiner, "Biden's CIA Pick Puts Spies Back in the Service of Statecraft," *Washington Post*, 12 January 2021.

20. Jeff Stein, "Bill Burns's CIA and the Roads not Taken," *Washington Monthly*, 1 April 2021.

21. *Washington Post*, 13 April 2021; Julian E. Barnes, "CIA Reorganization to Place New Focus on China," *New York Times*, 7 October 2021; Anton Troanovski and Julian E. Barnes, "US-Russian Engagement Deepens as CIA Chief Visits Moscow," *New York Times*, 3 November 2021; Morgan Phillips, "Biden CIA Head William Burns Says Pulling out of Afghanistan Will 'Diminish' US Intelligence," Fox News, 14 April 2021; *New York Times*, 17 April 2021; Thomas Gibbons-Neff and Julian E. Barnes, "Spy Agencies Seek New Afghan Allies as US Withdraws," *New York Times*, 14 May 2021.

22. Pew Research Center findings reported in Peter Wilby, "First Thoughts," *New Statesman*, 2 September 2021. Afghan public opinion was, however, a multi-toned, if not elusive beast. A Langer Research Associates poll published on 29 January 2015 recorded, "A record of 92 percent of Afghans prefer the current government over the Taliban, a sentiment that's been very widely held (by 82 to 92 percent) in nearly a decade of polling."

23. Burns, *Back Channel*, K302.

24. All quotations from the *Guardian*, 5 May 2021. Mija also stated, "I am a woman of color. I am a mom": *Washington Times*, 4 May 2021.

25. First quotation from title of editorial, "The Quiet Radical," *New Statesman*, 12–18 March 2021; second quotation from the title of Graham Greene's novel about the CIA's man in Saigon, first published in 1955; John O. Brennan quotations from his "Virtual Spy Chat" International Spy Museum Zoom presentation, 14 January 2021.
26. Remarks by President Biden at the office of the director of national intelligence, 27 July 2021.

CHAPTER 19

1. Panetta, *Worthy Fights*, 270, emphasis added.
2. "Committee Study," xxv.
3. See Lefever, *CIA and American Ethic*; Andregg, "Ethics and Professional Intelligence"; Omand and Phythian, *Principled Spying*; and Bellaby, *Ethics of Intelligence*.
4. Goldsmith, *Power and Constraint*, 87.
5. Gregory F. Treverton, email interview with the author, 16 April 2021.
6. Lowenthal, "Intelligence Officers," 12. Mark Lowenthal served in the CIA as an analyst and in the INR, worked for congressional intelligence committees, and became a widely published expert on intelligence matters.
7. Tenet quoting himself in Tenet, *Center of the Storm*, 68.

Bibliography

Adams, Sam. "Vietnam Cover-Up: Playing War with Numbers," *Harper's* (May 1975), 41–4, 62–73.

Agee, Philip. *Inside the Company: CIA Diary* (Harmondsworth: Penguin, 1975).

Ahern, Thomas L. *Vietnam Declassified: The CIA and Counterinsurgency* (Lexington, KY: University Press of Kentucky, 2010).

Aldrich, Richard J. "CIA History as a Cold War Battleground: The Forgotten First Wave of Agency Narratives," in Christopher R. Moran and Christopher J. Murphy, eds., *Intelligence Studies in Britain and the US: Historiography since 1945* (Edinburgh: Edinburgh University Press, 2013), 19–46.

"Alleged Assassination Plots Involving Foreign Leaders" (An Interim Report of the Select Committee to Study Governmental Operations with Respect to Intelligence Activities), *Senate Report*, 94 Cong., 1 sess. (20 November 1975).

Allen, George W. *None So Blind: A Personal Account of Intelligence Failure in Vietnam* (Chicago: Ivan R. Dee, 2001).

Alsop, Stewart. *The Center: The Anatomy of Power in Washington* (London: Hodder & Stoughton, 1968).

Alvarez, David, and Edouard Mark. *Spying through a Glass Darkly: American Espionage against the Soviet Union, 1945–1946* (Lawrence, KS: University Press of Kansas, 2016).

"American Republics: Cuba 1961–1962: Cuban Missile Crisis and Aftermath, Microfiche Supplement," ed. Edward C. Keefer, Louis J. Smith, and Charles S. Sampson, *FRUS*, 1961–3, vols. X–XII (1998).

Anderson, Martin. *Revolution* (San Diego, CA: Harcourt Brace Jovanovich, 1988).

Anderson, Scott. *The Quiet Americans* (New York: Doubleday, 2020).

Andregg, Michael. "Ethics and Professional Intelligence," in Loch K. Johnson, ed., *The Oxford Handbook of National Security Intelligence* (Oxford: Oxford University Press, 2010), 735–53.

Andrew, Christopher. *For the President's Eyes Only: Secret Intelligence and the American Presidency from Washington to Bush* (London: HarperCollins, 1995).

Andrew, Christopher. *The Secret World: A History of Intelligence* (New Haven, CT: Yale University Press, 2018).

Andrew, Christopher, and Oleg Gordievsky. *KGB: The Inside Story* (London: Hodder & Stoughton, 1990).

Angevine, Robert A. "Gentlemen Do Read Each Other's Mail: American Intelligence in the Interwar Era," *Intelligence and National Security*, 7 (April 1992), 1–29.

Arnold, Aaron, et al. "The Iran Nuclear Archive: Impressions and Implications," *Intelligence and National Security*, 36/2 (2021), 230–42.

Arnold, Jason Ross. *Secrecy in the Sunshine Era: The Promise and Failure of U.S. Open Government Laws* (Lawrence, KS: University Press of Kansas, 2014).

Auerbach, Jerold S. *Labor and Liberty: The La Follette Committee and the New Deal* (Indianapolis, IN: Bobbs-Merrill, 1966).

Bamford, James. *A Pretext for War* (New York: Random/Anchor, 2005).

Barnet, Richard J. "A Balance Sheet: Lippmann, Kennan, and the Cold War," in Michael J. Hogan, ed., *The End of the Cold War: Its Meanings and Implications* (Cambridge: Cambridge University Press, 1992), 113–27.

Barrett, David M. *The CIA and Congress: The Untold Story from Truman to Kennedy* (Lawrence, KS: University Press of Kansas, 2005).

Barrett, David M. "Retired Intelligence Officers versus President Trump," *Intelligence and National Security*, 35/1 (2020), 2–3.

Bearden, Milton A., and James Risen. *The Main Enemy: The Inside Story of the CIA's Final Showdown with the KGB* (New York: Random House, 2003).

Becker, Marc. *The FBI in Latin America: The Ecuador Files* (Durham, NC: Duke University Press, 2017).

Bellaby, Ross W. *The Ethics of Intelligence: A New Framework* (New York: Routledge, 2016).

Benjamin, Daniel, and Steven Simon. *The Age of Sacred Terror* (New York: Random House, 2002).

Bergen, Peter L. *Manhunt: The Ten-Year Search for Bin Laden from 9/11 to Abbottabad* (New York: Broadway, 2012).

Berkowitz, Bruce D. "US Intelligence Estimates of the Soviet Collapse: Reality and Perception," in *Ronald Reagan: Intelligence, and the End of the Cold War* (Simi Valley, CA: Ronald Reagan Presidential Library, 2011), 21–8.

Berle, Beatrice, and Travis B. Jacobs, eds. *Navigating the Rapids 1918–1971: From the Papers of Adolphe A. Berle* (New York: Harcourt Brace Jovanovich, 1973).

Best, Richard A., and Andrew Feickert. *Special Operations Forces (SOF) and CIA Paramilitary Operations: Issues for Congress* (Washington, D.C.: Congressional Research Service, 2006).

Bill, James A. *The Eagle and the Lion: The Tragedy of American-Iranian Relations* (New Haven, CT: Yale University Press, 1988).

Bird, Kay. *The Good Spy: The Life and Death of Robert Ames* (New York: Crown, 2014).

Blaufarb, Douglas S. *The Counterinsurgency Era: U.S. Doctrine and Performance, 1950 to the Present* (New York: Free Press, 1977).

Blight, James G., and Peter Kornbluh, eds. *Politics of Illusion: The Bay of Pigs Invasion Re-Examined* (Boulder, CO: Lynne Rienner, 1998).

Bohning, Don. *The Castro Obsession: U.S. Covert Operations against Cuba. 1959–1965* (Washington, D.C.: Potomac Books, 2005).

Booth, Alan R. "The Development of the Espionage Film," in Wesley K. Wark, ed., *Spy Fiction, Spy Films, and Real Intelligence* (London: Frank Cass, 1991), 136–60.

Bower, Tom. *Blind Eye to Murder: Britain, America and the Purging of Nazi Germany: A Pledge Betrayed* (London: Granada, 1984).

Bowles, Chester. *Promises to Keep: My Years in Public Life 1941–1969* (New York: Harper & Row, 1971).

Braden, Tom. "The Birth of the CIA," *American Heritage*, 28 (February 1977), 4–13.

Brennan, John. "Human Rights: A Case Study of Egypt" (MA thesis, University of Texas at Austin, 1980).

Brennan, John. *Undaunted: My Fight against America's Enemies* (New York: Caledon Books, 2020).

Breuer, William B. *Shadow Warriors: The Covert War in Korea* (New York: John Wiley, 1996).

Brown, Anthony Cave. *The Last Hero: Wild Bill Donovan* (New York: Times Books, 1982).

Brown, Archie. *The Human Factor: Gorbachev, Reagan, and Thatcher, and the End of the Cold War* (Oxford: Oxford University Press, 2020).

Bruce, David K. E. *OSS against the Reich: The World War II Diaries of David K. E. Bruce*, ed. Nelson D. Lankford (Kent, OH: Kent University Press, 1991).

Bryan, George S. *The Spy in America* (Philadelphia, PA: J. B. Lippincott, 1943).

Burns, Adam. *William Howard Taft and the Philippines: A Blueprint for Empire* (Knoxville, TN: University of Tennessee Press, 2020).

Burns, William J. *The Back Channel: American Diplomacy in a Disordered World* (London: Hurst, 2019).

Bush, George W. *Decision Points* (London: Virgin, 2010).

Byrne, Malcolm. *Iran-Contra: Reagan's Scandal and the Unchecked Abuse of Presidential Power* (Lawrence, KS: University Press of Kansas, 2014).

Cahn, Anne Hessing. *Killing Détente: The Right Attacks the CIA* (University Park, PA: Pennsylvania State University Press, 1998).

Cain, Frank. *The Australian Security Intelligence Organization: An Unofficial History* (Ilford: Frank Cass, 1994).

Callanan, James. *Covert Action in the Cold War: U.S. Policy, Intelligence, and CIA Operations* (New York: Tauris, 2010).

Campbell, Duncan. *The Unsinkable Aircraft Carrier: American Military Power in Britain* (London: Joseph, 1984).

Caute, David. *The Great Fear: The Anti-Communist Purge under Truman and Eisenhower* (London: Secker & Warburg, 1978).

Charles, Douglas M. "Before the Colonel Arrived: Hoover, Donovan, Roosevelt, and the Origins of American Central Intelligence," *Intelligence and National Security*, 20 (June 2005), 225–37.

Chester, Thomas. *Covert Network: Progressives, the International Rescue Committee, and the CIA* (Armonk, NY: M. E. Sharpe, 1995).

"China Deep Dive: The Intelligence Community's Capabilities and Competencies with Respect to the People's Republic of China," *Report of the House Permanent Committee on Intelligence*, 37-page undated redacted summary, released September 2020.

CIA: The Pike Report (London: Spokesman Books, 1977 [1976]).

Clapper, James R. *Facts and Fears* (New York: Viking, 2018).

Cline, Marjorie. *Teaching Intelligence in the Mid-1980s: A Survey of College and University Courses on the Subject of Intelligence* (Washington, D.C.: National Intelligence Study Center, 1985).

Cline, Ray S. *The CIA under Reagan, Bush and Casey: The Evolution of the Agency from Roosevelt to Reagan* (Washington, D.C.: Acropolis, 1981).

Clinton, William J. *My Life* (New York: Alfred A. Knopf, 2004).

Cockburn, Leslie. *Out of Control* (New York: Atlantic Monthly, 1987).

Coll, Steve. *Ghost Wars: The Secret History of the CIA, Afghanistan and Bin Laden* (New York: Penguin, 2004).

Coll, Steve. *Directorate S: The CIA and America's Secret Wars in Afghanistan and Pakistan* (New York: Penguin Press, 2018).

"Committee Study of the Central Intelligence Agency's Detention and Interrogation Program," *Report of the Senate Select Committee in Intelligence*, 113 Cong., 2 sess., 9 December 2014.

Conrad, Sherri. "Executive Order 12,333: Unleashing the CIA Violates the Leash Law," *Cornell Law Review*, 70 (June 1985), 968–90.

Corke, Sarah-Jane. *U.S. Covert Operations and Cold War Strategy: Truman, Secret Warfare and the CIA, 1945–53* (New York: Routledge, 2007).

Costello, John, and Oleg Tsarev. *Deadly Illusions: The KGB Orlov Dossier Reveals Stalin's Master Spy* (New York: Crown, 1993).

"Covert Action," *Hearings before the Select Committee with Respect to Intelligence Activities of the United States Senate*, 94 Cong., 7 vols., Vol I, 4 and 5 December 1975.

Crang, Jeremy. "Politics on Parade: Army Education and the 1945 General Election," *History*, 81 (April 1996), 215–27.

Cullather, Nick, ed. *Secret History: The CIA's Classified Account of Its Operations in Guatemala, 1952–1954*, 2nd edn (Stanford, CA: Stanford University Press, 2006 [1999]).

Cummings, Alfred. *Covert Action: Legislative Background and Possible Policy Questions* (Washington, D.C.: Congressional Research Service, 2007).

Curran, James. *Unholy Fury: Whitlam and Nixon at War* (Melbourne: Melbourne University Press, 2015).

Darling, Arthur B. *The Central Intelligence Agency: An Instrument of Government, to 1950* (University Park, PA: Pennsylvania State University Press, 1990 [1953]).

Davies, Philip H. J. *Intelligence and Government in Britain and the United States*, vol. 1: *Evolution of the U.S. Intelligence Community* (Santa Barbara, CA: Praeger, 2012).

Diamond, John. *The CIA and the Culture of Failure: US Intelligence from the End of the Cold War to the Invasion of Iraq* (Stanford, CA: Stanford University Press, 2008).

Dietl, Ralph L. *The Strategic Defense Initiative* (Lanham, MD: Lexington Books, 2018).

Dixon, Thomas Jr. *The Clansman: A Historical Romance of the Ku Klux Klan* (Ridgewood, NJ: Gregg Press, 1967 [1905]).

Dobbs, Michael. *One Minute to Midnight: Kennedy, Khrushchev, and Castro on the Brink of Nuclear War* (New York: Alfred A. Knopf, 2008).

Documents from the US Espionage Den, 58 vols. (Tehran: Intisharat-i-Danishjunya-i Piruy-i Khatt-i Imam, 1980–7).

Doel, Ronald E., and Allan A. Needell, "Science, Scientists, and the CIA: Balancing International Ideals, National Needs, and Professional Opportunities," in Rhodri Jeffreys-Jones and Christopher Andrew, eds., *Eternal Vigilance: Fifty Years of the CIA* (London: Frank Cass, 1997), 59–81.

Donovan, Michael M. "National Intelligence and the Iranian Revolution," in Rhodri Jeffreys-Jones and Christopher Andrew, eds., *Eternal Vigilance: Fifty Years of the CIA* (London: Frank Cass, 1997), 143–63.

Donovan, Michael M. "U.S. Political Intelligence and American Policy on Iran, 1950–1979" (Ph.D. dissertation, University of Edinburgh, 1997).

Downing, Taylor. *1983: The World at the Brink* (London: Little, Brown, 2018).

Doyle, Kate. "The End of Secrecy: US National Security and the Imperative for Openness," *World Policy Journal*, 16 (Spring 1999), 34–51.

Draper, Theodore. *A Very Thin Line: The Iran-Contra Affairs* (New York: Hill & Wang, 1991).

Dujmovic, Nicholas. "Ronald Reagan, Intelligence, William Casey, and CIA: A Reappraisal," in *Ronald Reagan, Intelligence, and the End of the Cold War* (Simi Valley, CA: Ronald Reagan Presidential Library, 2011), 8–20.

Durbin, Brent. *The CIA and the Politics of U.S. Intelligence Reform* (New York: Cambridge University Press, 2017).

Dylan, Huw, David V. Gioe, and Michael S. Goodman. *The CIA and the Pursuit of Security: History, Documents and Contexts* (Edinburgh: Edinburgh University Press, 2020).

"Eastern Europe: The Soviet Union," ed. Rogers P. Churchill and William Slany, *FRUS*, 1946, vol. VI (1969).

Ellsworth, Robert F., and Kenneth L. Adelman. "Foolish Intelligence," *Foreign Policy*, 36 (1979), 147–59.

Elwell-Sutton, Lawrence P. *Persian Oil: A Study in Power Politics* (London: Lawrence & Wishart, 1955).

"Emergence of the Intelligence Establishment," ed. C. Thomas Thorne, *FRUS*, 1945–50 (1996).

Epstein, Edward J. *Deception: The Invisible War between the KGB and the CIA* (New York: Simon & Schuster, 1989).

Erwin, Marshall C., and Amy Belasco. *Intelligence Spending and Appropriations: Issues for Congress* (Washington, D.C.: Congressional Research Service, 2013).

"Extraordinary Rendition in U.S. Counterterrorism Policy: The Impact on Transatlantic Relations," *Joint Hearing* before the Subcommittee on International Organizations, Human Rights, and Oversight and the Subcommittee on Europe of the Committee on Foreign Affairs, House of Representatives, 100 Cong., 1 sess., 17 April 2007.

Faddis, Sam. *The CIA War in Kurdistan: The Untold Story of the Northern Front in the Iraq War* (Philadelphia, PA: Casemate, 2020).

Fanon, Franz. *The Wretched of the Earth*, trans. Constance Farrington (Harmondsworth: Penguin, 1967 [1961]).

Farber, David. *Taken Hostage: The Iran Hostage Crisis and America's First Encounter with Radical Islam* (Princeton, NJ: Princeton University Press, 2005).

Ferris, John. "Coming in from the Cold War: The Historiography of American Intelligence, 1945–1990," *Diplomatic History*, 19 (Winter 1995), 87–115.

Final Report of the Select Committee to Study Governmental Operations with Respect to Intelligence Activities (Church Report), *Senate Report*, 94 Cong., 2 sess. (1976): Book 1: "Foreign and Military Intelligence"; Book 2: "Intelligence and the Rights of Americans"; Book 3: "Supplementary Detailed Staff Report on Intelligence Activities and the Rights of Americans"; Book 4: "Supplementary Detailed Staff Reports on Foreign and Military Intelligence"; Book 5: "The Investigation of the Assassination of President John F. Kennedy: Performance of the Intelligence Agencies."

Fingar, Thomas. *Reducing Uncertainty: Intelligence Analysis and National Security* (Stanford, CA: Stanford University Press, 2011).

Fingar, Thomas. *From Mandate to Blueprint: Lessons from Intelligence Reform* (Stanford, CA: Stanford University Press, 2021).

Finn, Peter, and Petra Couvée. *The Zhivago Affair: The Kremlin, the CIA, and the Battle over a Forbidden Book* (New York: Vintage, 2015).

Fletcher, Katy. "Evolution of the Modern American Spy Novel," *Journal of Contemporary History*, 22 (April 1987), 319–31.

Ford, Harold. *The CIA and the Vietnam Policymakers: Three Episodes, 1962–1968* (Washington, D.C.: Center for the Study of Intelligence, 1968).

Freedman, Lawrence. *US Intelligence and the Soviet Strategic Threat* (London: Macmillan, 1977).

Gaddis, John Lewis. *George F. Kennan: An American Life* (New York: W. W. Norton, 2011).

Gates, Robert M. *From the Shadows: The Ultimate Insider's Story of Five Presidents and How They Won the Cold War* (New York: Simon & Schuster, 1996).

Gates, Robert M. *Duty: Memoirs of a Secretary at War* (London: W. H. Allen, 2015).

Gati, Charles. *Failed Illusions: Moscow, Washington, Budapest and the 1956 Hungarian Revolt* (Stanford, CA: Stanford University Press, 2006).

Gentry, John A. "An INS Special Forum: US Intelligence Officers' Involvement in Political Activities in the Trump Era," *Intelligence and National Security*, 35/1 (2020), 1–19.

George, Roger Z. "Reflections on CIA Analysis: Is It Finished?" *Intelligence and National Security*, 26/1 (2011), 72–81.

Gerard, Emmanuel, and Bruce Kuklick. *Death in the Congo: Murdering Patrice Lumumba* (Cambridge, MA: Harvard University Press, 2015).

Ghazvinian, John. *America and Iran: A History, 1720 to the Present* (London: Oneworld, 2020).

Gleijeses, Piero. *Shattered Hope: The Guatemalan Revolution and the United States, 1944–1954* (Princeton, NJ: Princeton University Press, 1991).

Gleijeses, Piero. "Ships in the Night: The CIA, the White House and the Bay of Pigs," *Journal of Latin American Studies*, 27 (February 1995), 1–42.

Goldsmith, Jack. *Power and Constraint: The Accountable Presidency after 9/11* (New York: Norton, 2012).

Goncharov, German Arsenievich. "American and Soviet H-bomb Development Programmes: Historical Background," *Physics-Uspekhi*, 39/10 (1996), 1033–44.

Gooding, John. *Rulers and Subjects: Government and People in Russia 1801–1991* (London: Arnold, 1996).

Goodman, Allan E., Gregory F. Treverton, and Philip Zelikow, *In from the Cold* (New York: Twentieth Century Fund Press, 1996).

Goodman, Melvin A. "Ending the CIA's Cold War Legacy," *Foreign Policy*, 106 (Spring 1997), 128–43.

Goodman, Melvin A. *Failure of Intelligence: The Decline and Fall of the CIA* (Lanham, MD: Rowman and Littlefield, 2008).

Greene, Graham. *The Quiet American* (London: Heinemann, 1955).

Griffin, David Ray. *The 9/11 Commission Report: Omissions and Distortions* (Northampton, MA: Olive Press, 2005).

Gronke, Paul, et al. "US Public Opinion on Torture, 2001–2009," *PS: Political Science and Politics*, 443 (July 2010), 437–44.

Gustafson, Kristian, and Christopher Andrew. "The Other Hidden Hand: Soviet and Cuban Intelligence in Allende's Chile," *Intelligence and National Security*, 33/3 (2018): 407–21.

Hadley, David P. *The Rising Clamor: The American Press, the Central Intelligence Agency, and the Cold War* (Lexington, KY: University Press of Kentucky, 2019).

Harris Smith, Richard. *OSS: The Secret History of America's First Central Intelligence Agency* (Berkeley, CA: University of California Press, 1972).

Haslam, Jonathan. *The Nixon Administration and the Death of Allende's Chile: A Case of Assisted Suicide* (London: Verso, 2005).

Haslam, Jonathan. *Near and Distant Neighbors: A New History of Soviet Intelligence* (New York: Farrar, Straus and Giroux, 2015).

Hayden, Michael. *Playing to the Edge: American Intelligence in the Age of Terror* (New York: Penguin, 2016).

Haynes, John E., and E. Harvey Klehr, *Venona: Decoding Soviet Espionage in America* (New Haven, CT: Yale University Press, 1999).

Hedley, John H. "The Evolution of Intelligence Analysis," in Roger Z. George, ed., *Analyzing Intelligence: Origins, Obstacles, and Innovations* (Washington, D.C.: Georgetown University Press, 2008), 19–34.

Herbig, Katherine L., and Martin F. Wiskoff, *Espionage against the United States by American Citizens 1947–2001* (Monterey, CA: Defense Personnel Security Research Center, 2002).

Hodgson, Godfrey. *The Gentleman from New York: Daniel Patrick Moynihan: A Biography* (Boston, MA: Houghton Mifflin, 2000).

Hogan, Michael J., ed. *The End of the Cold War: Its Meanings and Implications* (Cambridge: Cambridge University Press, 1992).

Hogan, Michael J. *A Cross of Iron: Harry S. Truman and the Origins of the National Security State, 1945–1954* (Cambridge: Cambridge University Press, 1998).

Holtzman, Michael. *James Jesus Angleton, the CIA, and the Craft of Counterintelligence* (Amherst, MA: University of Massachusetts Press, 2008).

Hulnick, Arthur. *Keeping us Safe: Secret Intelligence and Homeland Security* (Westport, CT: Praeger, 2004).

Hutchings, Robert, and Gregory F. Treverton, eds. *Truth to Power: A History of the US National Security Council* (New York: Oxford University Press, 2019).

Immerman, Richard H. *The CIA in Guatemala: The Foreign Policy of Intervention* (Austin, TX: University of Texas Press, 1982).

Immerman, Richard H. *The Hidden Hand: A Brief History of the CIA* (Chichester: Wiley Blackwell, 2014).

Immerman, Richard H. "Tradecraft, the PIAB, and the 2007 NIE on Iran's Nuclear Intentions and Capabilities," *Intelligence and National Security*, 36/2 (2021), 217–21.

Immerman, Richard H. "Trump to the Intelligence Community: You're Fired," *H-Diplo/ISSF Policy Series 2021–4* (28 January 2021).

"Information Technology and Cyber Operations: Modernization and Policy Issues to Support the Future Force," *Hearing before the Subcommittee on Intelligence, Emerging Threats and Capabilities of the Committee on Armed Services*, 113 Cong., 1 sess., 13 March 2013.

"Iran, 1951–1954," ed. James C. Van Hook, *FRUS*, 1952–4 (2017).

"Iran: Hostage Crisis, November 1979–September 1980," ed. Linda Qaimmaqami, *FRUS*, 1977–80, vol. XI, pt. 1 (2020).

Jacobson, Adam D. "Back to the Dark Side: Explaining the CIA's Repeated Use of Torture," *Terrorism and Political Violence*, 33/2 (2021), 257–70.

Jeffreys-Jones, Rhodri. "The Socio-Educational Composition of the CIA Elite: A Statistical Note," *Journal of American Studies*, 19 (December, 1985), 421–4.

Jeffreys-Jones, Rhodri. *Peace Now! American Society and the Ending of the Vietnam War* (New Haven, CT: Yale University Press, 1999).

Jeffreys-Jones, Rhodri. "Man of the People? JFK and the Cuban Missile Crisis," *Reviews in American History*, 30 (September, 2002), 486–91.

Jeffreys-Jones, Rhodri. *The CIA and American Democracy*, 3rd edn (New Haven, CT: Yale University Press, 2003 [1989]).

Jeffreys-Jones, Rhodri. *In Spies We Trust: The Story of Western Intelligence* (Oxford: Oxford University Press, 2013).

Jeffreys-Jones, Rhodri. *The American Left: Its Impact on Politics and Society since 1900* (Edinburgh: Edinburgh University Press, 2013).

Jeffreys-Jones, Rhodri. *The Nazi Spy Ring in America: Hitler's Agents, the FBI, and the Case That Stirred a Nation* (Washington, D.C.: Georgetown University Press, 2020).

Jenkins, Tricia. *The CIA in Hollywood: How the Agency Shapes Film and Television* (Austin, TX: University of Texas Press, 2012).

Jensen, Steven L. B. *The Making of International Human Rights: The 1960s, Decolonization, and the Reconstruction of Global Values* (New York: Cambridge University Press, 2016).

Johnson, James W. *Along This Way: The Autobiography of James Weldon Johnson* (New York: Penguin, 1990 [1955]).

Johnson, Loch K., ed. *Strategic Intelligence*, vol. 4: *Counterintelligence and Counterterrorism: Defending the Nation against Hostile Forces* (Westport, CT: Praeger, 2007).

Johnson, Loch K. *The Threat on the Horizon: An Inside Account of America's Search for Security after the Cold War* (New York: Oxford University Press, 2011).

Johnson, Loch K. *National Security Intelligence* (Cambridge: Polity, 2012).

Johnson, Loch K. *Spy Watching: Intelligence Accountability in the United States* (New York: Oxford University Press, 2018).

Johnston, James H. *Murder, Inc.: The CIA under John F. Kennedy* (Lincoln, NE: Potomac/University of Nebraska Press, 2019).

Jones, Christopher M. "The CIA under Clinton: Continuity and Change," *International Journal of Intelligence and Counterintelligence*, 14/4 (2001), 503–28.

Jones, Howard. *The Bay of Pigs* (Oxford: Oxford University Press, 2008).

Jones, John P. *The German Spy in America: The Secret Plotting of German Spies in the United States and the Inside Story of the Sinking of the Lusitania* (London: Hutchinson, 1917).

Jones, Mathew. "Journalism, Intelligence and the *New York Times*: Cyrus L. Sulzberger, Harrison E. Salisbury and the CIA," *History*, 100 (April 2015), 229–50.

Jones, Nate. *Able Archer 83: The Secret History of the NATO Exercise That Almost Triggered Nuclear War* (New York: The New Press, 2016).

Kahn, David. *The Reader of Gentlemen's Mail: Herbert O. Yardley and the Birth of American Codebreaking* (New Haven, CT: Yale University Press, 2004).

Karalekas, Anne. "History of the Central Intelligence Agency," in William M. Leary, ed., *The Central Intelligence Agency: History and Documents* (Tuscaloosa, AL: University of Alabama Press, 1984).

Katz, Barry M. *Foreign Intelligence: Research and Analysis in the Office of Strategic Services 1942–1945* (Cambridge, MA: Harvard University Press, 1989).

Kearns, Erin M., and Joseph K. Young. *Tortured Logic: Why Some Americans Support the Use of Torture in Counterterrorism* (New York: Columbia University Press, 2020).

Keefe, Patrick R. "Privatized Spying: The Emerging Intelligence Industry," in Loch K. Johnson, ed., *The Oxford Handbook of National Security Intelligence* (Oxford: Oxford University Press, 2010), 296–309.

Kendall, Willmoore. "The Function of Intelligence," *World Politics*, 1 (1948–9), 542–52.

Kennan, George F. *American Diplomacy 1900–1950* (New York: New American Library, 1964 [1951]).

Kissinger, Henry. *White House Years* (Boston, MA: Little, Brown, 1979).

Klehr, Harvey, John E. Haynes, and Fridrikh I. Firsov, *The Secret World of American Communism* (New Haven, CT: Yale University Press, 1996).

Knott, Stephen F. *Secret and Sanctioned: Covert Operations and the American Presidency* (New York: Oxford University Press, 1996).

Kojm, Christopher. "Intelligence Integration and Reform: 2009–2014," in Robert Hutchings and Gregory F. Treverton, eds., *Truth to Power: A History of the US National Security Council* (New York: Oxford University Press, 2019), 157–97.

Kolko, Gabriel. *The Roots of American Foreign Policy: An Analysis of Power and Purpose* (Boston, MA: Beacon, 1969).

Kort, Michael. *The Soviet Colossus: A History of the USSR* (London: Routledge, 1992).

LaFeber, Walter. *America, Russia, and the Cold War, 1945–1996*, 8th edn (New York: McGraw-Hill, 1997 [1967]).

Lansing, Robert. *War Memoirs of Robert Lansing, Secretary of State* (Indianapolis, IN: Bobbs-Merrill, 1935).

Laqueur, Walter. *America, Russia, and the Cold War 1945–1971*, 2nd edn (New York: John Wiley, 1972).

Laville, Helen. "The Committee of Correspondence: CIA Funding of Women's Groups, 1952–1967," in Rhodri Jeffreys-Jones and Christopher Andrew, eds., *Eternal Vigilance? 50 Years of the CIA* (London: Frank Cass, 1997), 104–21.

Leary, William M., ed. *The Central Intelligence Agency: History and Documents* (Tuscaloosa, AL: University of Alabama Press, 1984).

Lederer, William J., and Eugene Burdick, *The Ugly American* (London: Victor Gollancz, 1959).

Lefever, Ernest W., and Roy Godson, *The CIA and the American Ethic: An Unfinished Debate* (Washington, D.C.: Ethics and Public Policy Center, Georgetown University, 1979).

Logevall, Fredrik. "Anatomy of an Unnecessary War: The Iraq Invasion," in Julian E. Zelizer, ed., *The Presidency of George Bush: A First Historical Assessment* (Princeton, NJ: Princeton University Press, 2010), 88–113.

Lorenz, Marita. *The Spy Who Loved Castro*, trans. Maria White (London: Ebury/Penguin, 2017 [2015]).

Lowenthal, Mark M. "Intelligence Officers, Oaths and Loyalty," *Intelligence and National Security*, 35/1 (2019), 11–12.

Lucas, Scott. *Freedom's War: The US Crusade against the Soviet Union* (Manchester: Manchester University Press, 1999).

McAdams, Dan P. *George W. Bush and the Redemptive Dream: A Psychological Portrait* (Oxford: Oxford University Press, 2011).

McAdams, Dan P. "Redemptive Narratives in the Life and Presidency of George W. Bush," in Charles B. Storzier, Daniel Offer, and Oliger Abdyli, eds., *The Leader: Psychological Essays*, 2nd edn (New York: Springer, 2011), 135–51.

McCarthy, David Shamus. *Selling the CIA: Public Relations and the Culture of Secrecy* (Lawrence, KS: University Press of Kansas, 2018).

McClintock, Michael. *Instruments of Statecraft: U.S. Guerrilla Warfare, Counterinsurgency, and Counterterrorism, 1940–1990* (New York: Pantheon, 1992).

McDonald, James K. "CIA and Warning Failures," in Lars C. Jenssen and Olave Riste, eds., *Intelligence in the Cold War: Organisation, Role and International Cooperation* (Oslo: Norwegian Institute for Defence Studies, 2001): 41–52.

McGarr, Paul. " 'Do We Still Need a CIA?' Daniel Patrick Moynihan, the Central Intelligence Agency and US Foreign Policy," *History*, 100 (April 2015), 275–92.

McGarvey, Patrick J. *CIA: The Myth and the Madness* (New York: Saturday Review Press, 1972).

Macintyre, Ben. *The Spy and the Traitor* (London: Penguin, 2018).

MacPherson, Nelson. *American Intelligence in War-Time London: The Story of OSS* (London: Frank Cass, 2003).

Mahoney, Richard D. *JFK: Ordeal in Africa* (New York: Oxford University Press, 1983).

Mainwaring, Sarah. "Division D: Operation Rubicon and the CIA's Secret SIGINT Empire," *Intelligence and National Security*, 35/5 (2020), 623–40.

Mangold, Tom. *Cold Warrior: James Jesus Angleton: The CIA's Master Spy Hunter* (New York: Simon & Schuster, 1991).

Mann, James. *George W. Bush* (New York: Times Books/Henry Holt, 2015).

Mansoor, Peter R. *Surge: My Journey with General David Petraeus and the Remaking of the Iraq War* (New Haven, CT: Yale University Press, 2013).

Marchetti, Victor, and John D. Marks. *The CIA and the Cult of Intelligence* (New York: Dell/Knopf, 1975 [1974]).

Marrin, Stephen. "Why Strategic Intelligence Analysis Has Limited Influence on American Foreign Policy," *Intelligence and National Security*, 32/6 (2017), 725–42.

Mason, Robert, and Iwan Morgan, eds. *Seeking a New Majority: The Republican Party and American Politics, 1960–1980* (Nashville, TN: Vanderbilt University Press, 2013).

Maugham, W. Somerset. *Collected Short Stories*, vol. 3 (Harmondsworth: Penguin, 1971). [The original and slightly different version of these factually based stories was published as *Ashenden* in 1928.]

May, Ernest R., and Philip Zelikow. *The Kennedy Tapes: Inside the White House during the Cuban Missile Crisis* (Cambridge, MA: Belknap Press of Harvard University Press, 2002 [1997]).

Mazzetti, Mark. *The Way of the Knife: The CIA, a Secret Army, and a War at the Ends of the Earth* (New York: Penguin, 2013).

Mendez, Antonio, and Jonna Mendez. *The Moscow Rules: The Secret CIA Tactics That Helped America Win the Cold War* (New York: Public Affairs, 2019).

Metzger, Fleur W. "Spies like Us?" *Denison Magazine* (Winter 1991), 12–17.

Milani, Abbas. *The Shah* (New York: Palgrave Macmillan, 2012).

Miles, Anne D. *Perspectives on the Senate Select Committee on Intelligence (SSCI) Study and Enhanced Interrogation Techniques* (Washington, D.C.: Congressional Research Service, 2015).

Miles, Simon. "The War Scare That Wasn't: Able Archer and the Myths of the Second Cold War," *Journal of Cold War Studies*, 22 (Summer 2020), 86–118.

Miller-Idris, Cynthia. *Hate in the Homeland: The New Global Far Right* (Princeton, NJ: Princeton University Press, 2020).

Mistry, Kaeten. "Approaches to Understanding the Inaugural Covert Operation in Italy: Exploding Useful Myths," *Intelligence and National Security*, 26/2–3 (2011), 246–68.

Mistry, Kaeten. "A Transnational Protest against the National Security State: Whistle-Blowing, Philip Agee, and Networks Dissent," *Journal of American History*, 106 (September 2019), 362–89.

Mistry, Kaeten, and Hannah Gurman. *Whistle-Blowing Nation: The History pf National Security Disclosures and the Cult of State Secrecy*. (New York: Columbia University Press, 2020.

Montague, Ludwell Lee. *General Walter Bedell Smith as Director of Central Intelligence October 1950–February 1953* (University Park, PA: Pennsylvania University Press, 1992 [1971]).

Moran, Christopher. *Company Confessions: Secrets, Memoirs and the CIA* (New York: St. Martin's Press, 2016).

Morell, Michael. *The Great War of Our Time: The CIA's Fight against Terrorism— From Al Qa'ida to ISIS* (New York: Hachette, 2015).

Morgan, Ted. *A Covert Life: Jay Lovestone: Communist. Anti-Communist, and Spymaster* (New York: Random House, 1999).

Morgan, Thomas B. *The Anti-Americans* (London: Michael Joseph, 1967).

Morris, Kenneth E. *Jimmy Carter: American Moralist* (Athens: University of Georgia Press, 1996).

Morrison, Lynne G.P. "Journalists and American Intelligence 1947 to 1955 with Particular Reference to the CIA" (MSc thesis, University of Edinburgh, 1991).

Mosley, Leonard. *Dulles: A Biography of Eleanor, Allen, and John Foster Dulles and Their Family Network* (London: Hodder & Stoughton, 1978).

Moyar, Mark. *Phoenix and the Birds of Prey: The CIA's Secret Campaign to Destroy the Viet Cong* (Annapolis, MD: Naval Institute Press, 1997).

Murphy, David E., Sergei Kondrashev, and George Bailey, *Battleground Berlin: CIA vs. KGB in the Cold War* (New Haven, CT: Yale University Press, 1997).

Naftali, Timothy. *Blind Spot: The Secret History of American Counterterrorism* (New York: Basic, 2005).

Naftali, Timothy. "George Bush and the 'War on Terror,'" in Julian E. Zelizer, ed., *The Presidency of George Bush: A First Historical Assessment* (Princeton, NJ: Princeton University Press, 2010), 59–87.

Naimark, Norman M. *Stalin and the Fate of Europe: The Postwar Struggle for Sovereignty* (Cambridge, MA: Belknap Press of Harvard University Press, 2019).

Nash, Philip. *The Other Missiles of October: Eisenhower, Kennedy, and the Jupiters, 1957–1963* (Chapel Hill, NC: North Carolina University Press, 2000).

National Security Council, "Foreign Threats to the 2020 US Federal Elections" (Intelligence Community Assessment ICA 2020-00078D, 10 March 2021).

Newman, David, and Tyll Van Geel. "Executive Order 12,333: The Risks of a Clear Declaration of Intent," *Harvard Journal of Law and Public Policy*, 12 (Spring 1989), 433–48.

Newman, John. *Oswald and the CIA* (New York: Carroll & Graf, 1995).

Nixon, Richard M. *Public Papers of the Presidents of the United States. Richard Nixon* (Washington, D.C.: GPO, n.d.).

Nixon, Richard M. *The Memoirs of Richard Nixon*, 2 vols. (New York: Warner, 1979).

Nolan, Bridget Rose. "Information Sharing and Collaboration in the United States Intelligence Community: An Ethnographic Study of the National Counterterrorism Center" (Ph.D. dissertation, University of Pennsylvania, 2013).

"Nomination of Robert M. Gates," *Hearings before the Select Committee on Intelligence of the United States Senate*, 100 Cong., 1 sess., 3 vols., vol. II, 24 September and 1, 2 October 1991.

Ochiai, Yukiko. "Vietnam Estimate: US Intelligence and the Origins of the Vietnam War, 1962–1965" (Ph.D. dissertation, University of Edinburgh, 2010).

Oliver, Willard M. *The Birth of the FBI: Teddy Roosevelt, the Secret Service, and the Fight over America's Premier Law Enforcement Agency* (Washington, D.C.: Rowman & Littlefield, 2019).

Olmsted, Kathryn S. *Challenging the Secret Government: The Post-Watergate Investigations of the CIA and FBI* (Chapel Hill, NC: University of North Carolina Press, 1996).

Olmsted, Kathryn S. *Real Enemies: Conspiracy Theories and American Democracy, World War I to 9/11* (Oxford: Oxford University Press, 2009).

Omand, David, and Mark Phythian. *Principled Spying: The Ethics of Secret Intelligence* (Oxford: Oxford University Press, 2018).

O'Rourke, Lindsay A. *Covert Regime Change: America's Secret Cold War* (Ithaca, NY: Cornell University Press, 2018).

Paget, Karen M. *Patriotic Betrayal: The Inside Story of the CIA's Secret Campaign to Enroll American Students in the Crusade against Communism* (New Haven, CT: Yale University Press, 2015).

Panetta, Leon. *Worthy Fights: A Memoir of Leadership in War and Peace* (New York: Penguin, 2014).

Petersen, Neal H. *American Intelligence, 1775–1990: A Bibliographic Guide* (Claremont, CA: Regina, 1992).

Pfeiffer, Jack B. "Official History of the Bay of Pigs Operation," 5 vols., vol. V, "CIA's Internal Investigation of the Bay of Pigs" (CIA, 18 April 1984 [released 2016]).

Phillips, David A. *The Night Watch: 25 Years of Peculiar Service* (New York: Atheneum, 1977).

Pipes, Richard. "Team B: The Reality Behind the Myth," *Commentary*, 82 (October 1986), 25–40.

Pisani, Sallie. *The CIA and the Marshall Plan* (Edinburgh: Edinburgh University Press, 1991).

Plokhy, Serhii. *Nuclear Folly: A New History of the Cuban Missile Crisis* (London: Allen Lane, 2021).

Portes, Jacques. *Les Américains et la guerre du Vietnam* (Brussels: Éditions Complexe, 1993).

Posner, Richard A. *Preventing Surprise Attacks: Intelligence Reform in the Wake of 9/11* (Stanford, CA: Hoover Institution, 2005).

Powers, Thomas. *The Man Who Kept the Secrets: Richard Helms and the CIA* (New York: Washington Square Press, 1981).

Prados, John. *The Soviet Estimate: US Intelligence Analysis and Russian Military Strength* (New York: Dial, 1982).

Prados, John. *Lost Crusader: The Secret Wars of CIA Director William Colby* (New York: Oxford University Press, 2003).

Prados, John. *Hoodwinked: The Documents That Reveal How Bush Sold Us a War* (New York: New Press, 2004).

Prados, John. *Safe for Democracy: The Secret Wars of the CIA* (Chicago: Ivan R. Dee, 2006).

Priest, Dana, and William M. Arkin. *Top Secret America: The Rise of the New American Security State* (New York: Little, Brown, 2011).

Puyvelde, Damien Van. "Quelle leçons tirer de la privatisation du renseignement aux États-Unis?" *Revue internationale et stratégique*, 87 (2012/13), 42–52.

Puyvelde, Damien Van. *Outsourcing US Intelligence: Contractors and Government Accountability* (Edinburgh: Edinburgh University Press, 2019).

Rahnema, Ali. *Behind the 1953 Coup in Iran: Thugs, Turncoats, Soldiers, and Spooks* (Cambridge: Cambridge University Press, 2015).

Ranelagh, John. *The Agency: The Rise and Decline of the CIA* (New York: Simon & Schuster, 1986).

Ransom, Harry H. "Secret Intelligence in the United States, 1947–1982: The CIA's Search for Legitimacy," in Christopher M. Andrew and David Dilks, eds., *The Missing Dimension: Governments and Intelligence Communities in the Twentieth Century* (Basingstoke: Macmillan, 1984), 199–226.

Rawick, George P., ed. *The American Slave: A Composite Autobiography*. vol. 2: *South Carolina Narratives* (Westport, CT: Greenwood, 1972).

Reagan, Ronald. *An American Life* (London: Hutchinson, 1990).

Reisman, M. Michael, and James E. Baker. *Regulating Covert Action: Practices, Contexts, and Politics of Covert Action Abroad in International and American Law* (New Haven, CT: Yale University Press, 1992).

"Review of the Terrorist Attacks on US Facilities in Benghazi, Libya, September 11–12, 2012" (by the US Senate Select Committee on Intelligence), *Senate Report*, 113 Cong., 2 sess., 15 January 2014.

Rice, Condoleezza. *No Higher Honor* (New York: Crown, 2011).

Risen, James. *State of War: The Secret History of the CIA and the Bush Administration* (New York: Free Press, 2006).

Robarge, David. "Moles, Defectors, and Deceptions: James Angleton and CIA Counterintelligence," *Journal of Intelligence History*, 3/2 (2003), 21–49.

Rollins, John. *Osama bin Laden's Death: Implications and Considerations* (Washington, D.C.: Congressional Research Service, 2011).

Ronge, Maximilian. *Kriegs- und Industrie-Spionage* (Leipzig: A. H. Payne, 1930).

Rovner, Joshua. "Preparing for a Nuclear Iran: The Role of the CIA," *Strategic Insights*, 4/11 (November 2005).

Rovner, Joshua. *Fixing the Facts: National Security and the Politics of Intelligence* (Ithaca, NY: Cornell University Press, 2011).

Rowan, Richard D. *The Story of Secret Service* (New York: Doubleday, Doran, 1937).

Rowe, Edward T. "The United States, the United Nations, and the Cold War," *International Organization*, 25 (Winter 1971), 59–78.

Roy, H. K. *American Spy: Wry Reflections on My Life in the CIA* (Amherst, NY: Prometheus Books, 2019).

Rudgers, David F. *Creating the Secret State: The Origins of the Central Intelligence Agency, 1943–1947* (Lawrence, KS: University Press of Kansas, 2000).

"Russian Active Measures Campaigns and Interference in the 2016 US Election. vol. 2: Russia's Use of Social Media," *Report of the [Senate] Select Committee on Intelligence*, 115 Cong., 1 sess., 10 November 2020.

Saunders, Frances Stonor. *Who Paid the Piper? The CIA and the Cultural Cold War* (London: Granta, 1999).

Scheuer, Michael. *Imperial Hubris: Why the West is Losing the War on Terror* (Washington, D.C.: Potomac, 2007 [2004]).

Scheuer, Michael. *Osama bin Laden* (Oxford: Oxford University Press, 2011).

Schlesinger, Arthur M., Jr. *The Age of Jackson* (Boston, MA: Little, Brown, 1945).

Schmitt, Michael N. "State-Sponsored Assassination in International and Domestic Law," *Yale Journal of International Law*, 17 (Summer 1992), 609–86.

Schorr, Daniel. *Clearing the Air* (Boston, MA: Houghton Mifflin, 1977).

Schroeder, Richard E. *Harry Truman, the Missouri Gang, and the Origins of the Cold War* (Columbia, MO: University of Missouri Press, 2017).

Scott, Len. "The 'Incredible Wrongness' of Nikita Khrushchev: The CIA and the Cuban Missile Crisis," *History*, 100 (2015), 210–28.

Scott, Len. "November 1983: The Most Dangerous Moment of the Cold War?" *Intelligence and National Security*, 35/1 (2019), 131–48.

Scott-Smith, Giles. *The Politics of Apolitical Culture: The Congress for Cultural Freedom, the CIA, and Post-War American Hegemony* (London: Routledge, 2002).

Serafino, Nina, Curt Tarnoff, and Dick K. Nanto. *US Occupation Assistance: Iraq, Germany and Japan Compared* (Washington, D.C.: Congressional Research Service, 2006).

Service, Robert. *The History of Twentieth Century Russia* (Harmondsworth: Penguin, 1998).

Sharp, U. S. Grant. *Strategy for Defeat: Vietnam in Retrospect* (San Rafael, CA: Presidio Press, 1982).

Shultz, George P. *Turmoil and Triumph: My Years as Secretary of State* (New York: Scribner's, 1993).

Sibley, Katherine A. S. *Red Spies in America: Stolen Secrets and the Dawn of the Cold War* (Lawrence, KS: University Press of Kansas, 2004).

Smist, Frank J. *Congress Oversees the United States Intelligence Community, 1947–1989* (Knoxville, TN: University of Tennessee Press, 1990).

Smith, Gerard. *Doubletalk: The Story of the First Strategic Arms Limitation Talks* (New York: Doubleday, 1980).

Smith, Joseph B. *Portrait of a Cold Warrior* (New York: Putnam, 1976).

Smith, Margaret Chase. *Declaration of Conscience* (Garden City, NY: Doubleday, 1972).

"Soviet Union, January 1983–March 1985," ed. Elizabeth C. Charles, *FRUS*, 1981–8, vol. IV (2021).

"Soviet Union, March 1985–October 1986," ed. Elizabeth C. Charles, *FRUS*, 1981–8, vol. V (2020).

Stafford, David. *Roosevelt and Churchill: Men of Secrets* (London: Little, Brown, 1999).

Stafford, David. *Spies beneath Berlin* (London: John Murray, 2002).

"START 1," ed. James G. Wilson, *FRUS*, 1981–8, vol. XI (2021).

Stern, Sol. "NSA and CIA," *Ramparts*, 5/9 (March 1967), 29–39.

Stewart, Gordon N. *The Cloak and Dollar War* (London: Lawrence & Wishart, 1953).

Stout, Mark. "World War I and the Birth of American Intelligence Culture," *Intelligence and National Security*, 32/3 (2017), 378–94.

Summers, Harry G., Jr. *On Strategy: A Critical Analysis of the Vietnam War* (Novato, CA: Presidio Press, 1982).

Talbot, David. *The Devil's Chessboard: Allen Dulles, the CIA, and the Rise of America's Secret Government* (New York: HarperCollins, 2015).

Tenet, George J. *At the Center of the Storm: My Years at the CIA* (New York: HarperCollins, 2007).

The 9/11 Report: Final Report of the National Commission on Terrorist Attacks upon the United States (New York: Norton, 2004).

"The Berlin Tunnel Operation 1952–1956" (CIA Clandestine Services History, 1967).

The Kennedy Presidential Press Conferences (London: Heyden, 1978).

The Making of US Policy in Iran, 1977–1980 (Alexandria, VA: Chadwyck-Healey for the National Security Archives, 1989).

Thomas, Evan. *The Very Best Men: Four Who Dared: The Early Years of the CIA* (New York: Touchstone, 1995).

Thompson, Robert. *Make for the Hills: Memories of Far Eastern Wars* (London: Leo Cooper, 1989).

Townley, Dafydd. "Too Responsible to Run for President: Frank Church and the 1976 Presidential Election," *Journal of Intelligence History*, published online 5 November 2020.

Trenta, Luca. "'An Act of Insanity and National Humiliation': The Ford Administration, Congressional Inquiries and the Ban on Assassination," *Journal of Intelligence History* 17/2 (2018): 121–40.

Truman, Harry S. *Memoirs*. I: *Year of Decisions* (Garden City, NY: Doubleday, 1955).

U.S. Commission on National Security/21st Century, *Road Map for Security: Imperative for Change* (Washington, D.C.: US Government Printing Office, 2001).

Vogel, Steve. *Betrayal in Berlin: George Blake, the Berlin Tunnel and the Greatest Conspiracy of the Cold War* (London: John Murray, 2019).

Waller, Douglas. *Wild Bill Donovan* (New York: Free Press, 2011).

Walsh, Lawrence E. *Iran-Contra: The Final Report* (New York: Times Books/ Random House, 1994).

Walton, Calder. "Spies, Election Meddling, and Disinformation: Past and Present," *Brown Journal of World Affairs*, 26 (Fall/Winter 2019), 107–24.

Wark, Wesley K. *The Ultimate Enemy: British Intelligence and Nazi Germany, 1933–1939* (Oxford: Oxford University Press, 1985).

Warner, Michael, ed. *Cold War Records: The CIA under Harry Truman* (Washington, D.C.: Center for the Study of Intelligence, CIA, 1994).

Waters, Robert A., and Gordon O. Daniels. "Striking for Freedom? International Intervention and the Guianese Sugar Workers' Strike of 1964," *Cold War History*, 10/4 (2010), 537–69.

Wegener, Jens. "Order and Chaos: The CIA's HYDRA Database and the Dawn of the Information Age," *Journal of Intelligence History*, 19/1 (2020), 77–91.

Weiner, Tim. *Legacy of Ashes: The History of the CIA* (London: Allen Lane/ Penguin, 2007).

Weisbrot, Robert. *Maximum Danger: Kennedy, the Missiles, and the Crisis of American Confidence* (Chicago: Ivan Dee, 2001).

West, Rebecca. *The New Meaning of Treason* (Harmondsworth: Penguin, 1981 [1949, 1964]).

Westwood, John N. *Endurance and Endeavour: Russian History, 1812–1992*, 4th edn (Oxford: Oxford University Press, 1993).

Whipple, Chris. *The Spymasters: How the CIA Directors Shape History and the Future* (New York: Simon & Schuster, 2020).

White, Mark J. *Missiles in Cuba: Kennedy, Khrushchev, Castro and the 1962 Crisis* (Chicago: Ivan R. Dee, 1997).

Wiebes, Cees. *Intelligence and the War in Bosnia, 1992–1995* (Münster: Lit Verlag, 2003).

Wilber, Donald N. "Overthrow of Premier Mossadeq of Iran: November 1952–August 1953," (CIA, Clandestine Services History, March 1954).

Wilford, Hugh. *The CIA, the British Left and the Cold War: Calling the Tune?* (London: Frank Cass, 2003).

Wilford, Hugh. *The Mighty Wurlitzer: How the CIA Played America* (Cambridge, MA: Harvard University Press, 2008).

Wilford, Hugh. *America's Great Game: The CIA's Secret Arabists and the Shaping of the Modern Middle East* (New York: Basic, 2014).

Wilford, Hugh. *The Agency: A History of the CIA: Course Guidebook* (Chantilly, VA: Teaching Co., 2019).

Wilkie, John E. "The Secret Service in the War," in *The American-Spanish War: A History by the War Leaders* (Norwich, CT: Chas. C. Haskell, 1899).

Williams, Grover S., ed. *Legislative History of the Central Intelligence Agency as Documented in Published Congressional Sources* (Washington, D.C.: Congressional Research Service, 1975).

Williams, William Appleman. *The Tragedy of American Diplomacy* (Cleveland, OH: World Publishing, 1959).

Willmetts, Simon. *In Secrecy's Shadow: The OSS and CIA in Hollywood Cinema 1941–1979* (Edinburgh: Edinburgh University Press, 2016).

Winks, Robin. *Cloak and Gown: Scholars in the Secret War, 1939–1961* (New York: Morrow, 1987).

Wise, David. *Nightmover: How Aldrich Ames Sold the CIA to the KGB for $4.6 Million* (New York: HarperCollins, 1995).

Woods, Randall B. *Shadow Warrior: William Egan Colby and the CIA* (New York: Basic, 2013).

Woodward, Bob. *Veil: The Secret Wars of the CIA, 1981–97* (London: Simon & Schuster, 1987).

Wriston, Henry M. *Executive Agents in American Foreign Relations* (Baltimore, MD: Johns Hopkins Press, 1929).

Wyden, Peter. *Bay of Pigs: The Untold Story* (London: Jonathan Cape, 1979).

Zelikow, Philip. "American Economic Intelligence: Past Practice and Future Principles," in Rhodri Jeffreys-Jones and Christopher Andrew, eds., *Eternal Vigilance: Fifty Years of the CIA* (London: Frank Cass, 1997), 178–200.

Index